Cambridge Studies in Biological and Evolut

Western Diseases

As a group, western diseases such as type 2 diabetes, cardiovascular disease, breast cancer, allergies and mental health problems constitute one of the major problems facing humans at the beginning of the twenty-first century, particularly as they extend into the poorer countries of the world. An evolutionary perspective has much to offer standard biomedical understanding of western diseases. At the heart of this approach is the notion that human evolution occurred in circumstances very different from the modern affluent western environment and that, as a consequence, human biology is not adapted to the contemporary western environment. Pollard provides a novel synthesis of this evolutionary perspective in a book aimed at advanced students and academics in the fields of medicine, human population biology and medical anthropology.

TESSA POLLARD graduated from the University of Oxford with degrees in Human Sciences and Biological Anthropology. She is currently a lecturer in Medical Anthropology at Durham University. She conducts research on risk factors for cardiovascular disease and type 2 diabetes in western and westernising populations.

Cambridge Studies in Biological and Evolutionary Anthropology

Series editors

HUMAN ECOLOGY
C. G. Nicholas Mascie-Taylor, University of Cambridge
Michael A. Little, State University of New York, Binghamton
GENETICS
Kenneth M. Weiss, Pennsylvania State University
HUMAN EVOLUTION
Robert A. Foley, University of Cambridge
Nina G. Jablonski, California Academy of Science
PRIMATOLOGY
Karen B. Strier, University of Wisconsin, Madison

Also available in the series

39 *Methods in Human Growth Research* Roland C. Hauspie, Noel Cameron & Luciano Molinari (eds.) 0 521 82050 2

40 *Shaping Primate Evolution* Fred Anapol, Rebecca L. German & Nina G. Jablonski (eds.) 0 521 81107 4

41 *Macaque Societies – A Model for the Study of Social Organization* Bernard Thierry, Mewa Singh & Werner Kaumanns (eds.) 0 521 81847 8

42 *Simulating Human Origins and Evolution* Ken Wessen 0 521 84399 5

43 *Bioarchaeology of Southeast Asia* Marc Oxenham & Nancy Tayles (eds.) 0 521 82580 6

44 *Seasonality in Primates* Diane K. Brockman & Carel P. van Schaik 0 521 82069 3

45 *Human Biology of Afro-Caribbean Populations* Lorena Madrigal 0 521 81931 8

46 *Primate and Human Evolution* Susan Cachel 0 521 82942 9

47 *The First Boat People* Steve Webb 0 521 85656 6

48 *Feeding Ecology in Apes and Other Primates* Gottfried Hohmann, Martha Robbins & Christophe Boesch (eds.) 0 521 85837 2

49 *Measuring Stress in Humans: A Practical Guide for the Field* Gillian Ice & Gary James (eds.) 0 521 84479 7

50 *The Bioarchaeology of Children: Perspectives from Biological and Forensic Anthropology* Mary Lewis 0 521 83602 6

51 *Monkeys of the Taï Forest* W. Scott McGraw, Klaus Zuberbühler & Ronald Noë (eds.) 0 521 81633 5

52 *Health Change in the Asia-Pacific Region: Biocultural and Epidemiological Approaches* Ryutaro Ohtsuka & Stanley J. Ulijaszek (eds.) 978 0 521 83792 7

53 *Technique and Application in Dental Anthropology* Joel D. Irish & Greg C. Nelson (eds.) 978 0 521 870 610

54 *Western Diseases: An Evolutionary Perspective* Tessa M. Pollard 978 0 521 61737 6

Western Diseases

An evolutionary perspective

Tessa M. Pollard
Durham University

CAMBRIDGE
UNIVERSITY PRESS

CAMBRIDGE UNIVERSITY PRESS
Cambridge, New York, Melbourne, Madrid, Cape Town, Singapore, São Paulo, Delhi

Cambridge University Press
The Edinburgh Building, Cambridge CB2 8RU, UK

Published in the United States of America by Cambridge University Press,
New York

www.cambridge.org
Information on this title: www.cambridge.org/9780521851800

© T. M. Pollard 2008

First published 2008

Printed in the United Kingdom at the University Press, Cambridge

A catalogue record for this publication is available from the British Library

ISBN 978-0-521-85180-0 hardback
ISBN 978-0-521-61737-6 paperback

Dedicated to
Geoffrey Ainsworth Harrison

Contents

Preface   *page xi*

1 Introduction   1
2 An evolutionary history of human disease   9
3 Obesity, type 2 diabetes and cardiovascular disease   23
4 The thrifty genotype versus thrifty phenotype debate:
   efforts to explain between population variation in rates
   of type 2 diabetes and cardiovascular disease   50
5 Reproductive cancers   75
6 Reproductive function, breastfeeding and the
   menopause   99
7 Asthma and allergic disease   120
8 Depression and stress   136
9 Conclusion   153

   References   173
   Index   217

ix

Preface

The inspiration for this book originates with Professor Geoffrey Ainsworth Harrison of the University of Oxford, who taught me and many others the value of the evolutionary approach to human biology. The third edition of the textbook he wrote with Paul Baker and others (Harrison *et al.* 1988) was a defining part of the curriculum at Oxford and I draw strongly on the approach of that volume in this book. He also introduced us to literature on western diseases published in the 1970s and 1980s by Boyden, and by Trowell and Burkitt, described at the beginning of Chapter 1. His supervision of my postgraduate work provided invaluable further opportunity to learn from his methods and ideas. In my subsequent career I have benefited greatly from this solid and stimulating foundation.

In my own teaching of advanced undergraduates and graduate students at Durham University I have felt the lack of an up-to-date equivalent of these texts, a feeling that was the main motivation for me to write this book. In the intervening years I have also benefited from exposure to epidemiological research on cardiovascular disease and other western diseases, and I hope that the end result profits from my learning beyond anthropology. My aim has been to draw these two approaches together to create a new synthesis. I am aware, of course, that in aiming for such a synthesis I have failed to provide the level of evolutionary theory that some evolutionary biologists and biological anthropologists would wish to see, while offering less detail of biomedical and epidemiological research than might be expected from the other side.

Many people provided practical and other kinds of support during the writing process and I would like to thank them. Alejandra Núñez-de la Mora commented on drafts of several chapters and I am particularly grateful for her time and effort. Thanks also to others who gave me valuable feedback and assistance. They are Gillian Bentley, Nigel Unwin, Ernie Pollard, Malcolm Smith, Grażyna Jasieńska, Malia Fullerton, Helen Ball, Peter Ellison, Judith Manghan, Andrew Russell, Leslie Carlin, Tim Gage, Robert Hegele, Bob Layton, Steve McGarvey, Trudi Buck, Caroline Jones, Peter Collins, Veryan Pollard and Kate Hampshire. I have also benefited greatly from interactions with students at Durham University. My greatest debt is to Malcolm Smith and Rosa Pollard Smith, who provided practical and emotional support, and showed patience and forbearance throughout. I am very grateful to them.

1 *Introduction*

This book sets out to examine why certain non-communicable diseases have become common, first in affluent western populations and now, increasingly, worldwide. I use an evolutionary perspective because of its value in showing us why and how human bodies are vulnerable to these diseases. In this chapter I introduce the concept of western diseases and outline the evolutionary perspective applied throughout the book.

Western diseases

In the 1960s and 1970s concerns developed about the rise of diseases such as coronary heart disease, type 2 diabetes and colon and breast cancer as important causes of mortality and morbidity in the western world (Cleave *et al.* 1969; Boyden 1970; Burkitt 1973). The origins of the term western diseases, and of the approach I adopt in this book, lie in this work. The diseases identified were linked to 'modern western civilisation' and were considered to be 'man-made' (Trowell and Burkitt 1981a). Specifically, the rise of western diseases was blamed on increased availability of food (accompanied by a decline in the consumption of dietary fibre) and a reduction in physical activity. These authors also acknowledged the impact of an increase in life expectancy, which led to a higher proportion of susceptible older people in the population. The emergence of western diseases in non-western societies, for example in the Far East and in Africa, was linked to the process of westernisation, that is, the adoption of elements of the modern western lifestyle in other areas of the world (Trowell and Burkitt 1981b), a simplistic but nevertheless helpful concept.

The most obvious alternative descriptors for western diseases are 'non-communicable diseases' and 'diseases of affluence', both of which I also employ, but in general I find them less useful in this context. 'Non-communicable diseases' includes diseases that are not associated with a western lifestyle, such as chronic obstructive pulmonary disease, common both in rich and poorer countries. 'Diseases of affluence' is unsatisfactory principally in that there is a difference between the disease profile of affluent countries in the 'west' and in the east. In particular, diets in the affluent east (e.g. in Japan

1

and South Korea) have traditionally been low in fat, and rates of coronary heart disease and other diseases strongly associated with dietary fat intake are, accordingly, lower than in the west (Ueshima *et al.* 2003). The second issue is that 'diseases of affluence' are not necessarily most prevalent among the most affluent members of a population. Within the United Kingdom, United States and many other countries, obesity and cardiovascular disease are more common in the poorer members of society (Marmot *et al.* 1984). High rates of type 2 diabetes and cardiovascular disease, in particular, are also increasingly common in poorer countries (Ezzati *et al.* 2005).

I have therefore chosen to retain the term western diseases, despite the fact that it is not geographically accurate, encompassing populations living as far apart as north America, western Europe and Australia. Its main advantages are that it neatly encapsulates the group of diseases with which I am concerned, and links to a paradigm, as described above, that I find helpful. The group of diseases I consider in this book is much as characterised above, although I also extend it to include allergies and mental health problems, as well as aspects of reproductive function and events, as detailed below.

An evolutionary perspective

This book draws on an evolutionary perspective to consider the rise of western diseases. As indicated by the recent surge of interest in evolutionary medicine (also called Darwinian medicine), an evolutionary perspective has much to offer the traditional understanding of biomedical science (Nesse and Williams 1994; Trevathan *et al.* 1999; Stearns and Ebert 2001; Nesse *et al.* 2006; Trevathan *et al.* 2007). At the heart of this approach is the notion that human evolution occurred in circumstances very different from the modern affluent western environment and that, as a consequence, human biology is not adapted to the contemporary western environment (Harrison 1973). It is worth examining this idea in more detail at the outset.

The precise evolutionary history of hominins (humans and their immediate ancestors) has been the subject of hotly contended debate for many years. Given the indirect and often sparse nature of the evidence, it is unlikely that a firm consensus will emerge for some time. Nevertheless, it is generally agreed that the human line diverged from the line leading to modern chimpanzees between about 5 and 7 million years ago (Boyd and Silk 2006). Our genus, *Homo*, appeared in Africa about two million years ago (Wood and Collard 1999), and subsequently these early hominins spread out of Africa, as far away as, for example, Indonesia. The emergence of *Homo* marked the first appearance of some important skeletal characteristics shared by all future members of the genus, including modern *Homo sapiens*. These features include large relative brain sizes, large bodies, dedicated bipedal locomotion and smaller teeth and jaws (Aiello and Wells 2002), features that were probably accompanied by marked changes in physiology and behaviour. For example, it is now thought that the energetic needs of *Homo*'s larger body size relative to earlier hominins and other primates were satisfied by an increase in the consumption of meat relative to other foods (Aiello and Wheeler 1995). The metabolic costs of the relatively large and energy-expensive human brain are also thought to have been met through a variety of characteristically human features. These include a corresponding reduction in the size of the equally energy-expensive gut (Aiello and Wheeler 1995) and a slow growth rate (Foley and Lee 1991). It is also likely that there was an increase in the relative proportion of adipose tissue (fat) at this time, as discussed further in Chapter 3.

The timing and location of the origins of our own species, anatomically modern *Homo sapiens*, has been a particular focus of debate in recent years, but the hypothesis that *Homo sapiens* first emerged in Africa about 100 000 years ago, often known as the 'Out of Africa' theory, is now the best supported (Stringer 2002; Stringer 2003). Subsequently, these modern humans colonised other parts of the world, where they may have entirely replaced the original hominin occupants, or may have interbred with them (Stringer 2003) (Fig. 1.1). In the last 50 000 years or so modern humans colonised Europe, Asia, the Pacific and eventually, around 20 000 years ago, they reached the Americas (Cavalli-Sforza and Feldman 2003). During the period in which

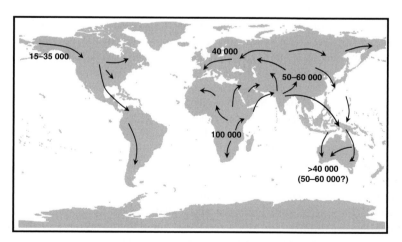

Fig. 1.1. Map showing the migration of modern *Homo sapiens* from Africa to the rest of the world. Figures are years before present. Redrawn with permission from Cavalli-Sforza and Feldman (2003).

different populations of anatomically modern *Homo sapiens* have lived in widely differing environments, they have experienced different selective pressures, so that we should expect some adaptive genetic differentiation to have arisen between populations, in addition to random differentiation due to isolation and the chance process of genetic drift (Tishkoff and Kidd 2004). This is important because such processes may have resulted in genetic differentiation between populations that affects their relative vulnerability to some western diseases.

The environment experienced by *Homo* from its emergence about two million years ago until the origins of agriculture about 10 000 years ago, the Palaeolithic period has been called the environment of evolutionary adaptedness (EEA) for humans (Bowlby 1969). During the Palaeolithic, hominins used stone tools and subsisted by a combination of hunting and gathering (also known as foraging). The EEA approach sees human biology as having adapted, through the process of natural selection, to the environment experienced during this long period. Individuals whose genes resulted in greater survival and fertility in the EEA, and which conferred those same advantages on their descendants, would have contributed more offspring to subsequent generations than those who were less well adapted. This greater reproductive success increased the relative frequency of genes conferring adaptation to that environment. This view is summarised by the notion that 'Human biology is designed for Stone Age conditions' (Williams and Nesse 1991). Some versions of this approach suggest that there has not been enough time for further evolutionary change since the origins of agriculture.

The concept of an EEA is a powerful one, but it has been criticised on a number of counts. Perhaps the most important criticism is of the implication that human evolution started two million years ago and ended ten thousand years ago. Clearly, this is not the case (Strassmann and Dunbar 1999). It is also likely that early species of *Homo*, who all lived within the last two million years and thus in the EEA as defined above, were very different, and lived in very different ways, from modern *Homo sapiens*. Certainly brain size increased markedly during this period. Nor was the environment static; there was major climatic change associated with glaciation (Foley 1996). To focus on the fact that humans were hunter–gatherers or foragers for one or two million years therefore creates a false picture of stasis in hominins and their environment during this period (Irons 1998). The approach adopted in this book is to follow Williams and Nesse and others in emphasising the importance of human adaptedness to the hunter–gatherer way of life, as experienced by modern humans before the origin of agriculture, whilst acknowledging that in discussing some traits it is important to draw on a deeper or shallower view of human evolutionary heritage.

Of course, an individual's biology is not determined solely by the genes he or she inherits as a consequence of past evolutionary processes. The environment he or she experiences helps determine his or her phenotype. Of particular relevance here, individuals are able to respond adaptively at the phenotypic level to different environments, and humans are good at this in comparison to other species (Thomas 1998). Such adaptation may occur as relatively short term and reversible acclimatisation, or by developmental adaptation, when an individual develops in a particular way in response to the environment experienced in early life. The long developmental period of humans probably contributes to our ability to adapt to prevailing environmental conditions in ways that improve future survival and reproduction (Vitzthum 2001). Biological anthropologists were among the first to recognise the importance of this developmental plasticity and of viewing an individual's biology as the outcome not only of his or her genetic inheritance, but also of his or her experiences over a lifetime, from life in the womb onwards (Lasker 1969). More recently, this lifespan perspective has become a focus of great interest in epidemiology (Barker 1994; Kuh and Hardy 2002; Kuh *et al.* 2003).

Of course, the ability to respond adaptively to the environment (and the limits to that adaptability) evolved over our long evolutionary history. There is little reason to expect an individual raised in an affluent western environment to adapt entirely successfully to his or her environment, as many aspects of it have never previously been encountered by humans. Furthermore, a rapidly changing environment poses special challenges to developmental adaptations. Permanent developmental adaptations are likely to be beneficial only when environments do not change a great deal during an individual's lifetime, so that information available during development serves as a reliable predictor of future conditions. During the relatively stable environments encountered during most of human evolution, such mechanisms would have worked well. However, if people encounter very different environmental conditions in adulthood from those during development, the developmental adjustments they make may not be beneficial. Today, this is particularly likely for populations in economic transition in the poorer countries of the world, and for migrants from poorer to wealthier circumstances, as we shall see.

We must also recognise that not all plasticity is adaptive. For example, lack of food may limit growth, and while some have argued that this is adaptive, most would now disagree. Often the task of distinguishing adaptive responses from those that arise simply as a result of environmental constraint is a difficult one (Ellison and Jasiénska 2007).

This, then, is the basis of the evolutionary perspective that I apply to this investigation of western diseases. The central point is that human biology is the product of a long evolutionary history in very different environments from the modern western environment, which is very new in evolutionary terms.

Outline of the book

In Chapter 2 I provide an evolutionary history of the human disease experience, starting with a consideration of diseases experienced by hunter–gatherers. My aim here is to provide important background information on the disease context in which modern *Homo sapiens* lived for the majority of the time since it emerged as a species. Next I examine the impact of the development of agriculture and of urban societies on human disease, innovations that exposed humans to more infectious disease. I then explore the concept of the epidemiologic transition, the term used to describe the decline in mortality from infectious disease and increase in mortality from non-communicable disease that started in the west after the industrial revolution. Finally, I summarise the position with respect to trends in the prevalence of western diseases over the last 50 years or so.

When western diseases were first identified as a group, they were poorly understood. It is now possible to draw on a huge wealth of information to examine each group of diseases, highlighting where an evolutionary perspective provides insights unavailable to standard biomedical science. In the main body of the book, from Chapter 3 to Chapter 8, I have aimed to do this for four major groups of diseases or health issues.

Obesity is the central pathology at the heart of many western diseases, and it plays a role in most others. Chapter 3 starts by reviewing research by biological anthropologists on diet and physical activity patterns in hunter–gatherer societies that helps us understand the energetic regimes to which our bodies are adapted. It then examines the reasons for the current increases in rates of obesity, and the most clearly obesity-related diseases, particularly type 2 diabetes and cardiovascular disease. Chapter 4 goes on to describe and critique two dominant hypotheses, often known as the thrifty genotype and thrifty phenotype hypotheses, that have been put forward to explain why some populations appear to be particularly susceptible to the development of type 2 diabetes and cardiovascular disease.

Breast cancer and other reproductive cancers in women are linked to exposure to oestrogen, and probably progesterone, produced by a woman's own body. In men, prostate cancer has been linked to high levels of testosterone. Research has shown that the high levels of these gonadal hormones in women and men in western societies today are an evolutionarily new phenomenon. This research is discussed in Chapter 5. Chapter 6 continues the reproductive health theme, but focuses mainly on women, examining the impact of a modern western lifestyle on reproductive function, infant feeding choices and their health consequences, and on the experience of the menopause.

In recent decades there has been an enormous rise in the prevalence of asthma and other allergic disease in children, particularly in the west, so that they can also be considered as western diseases. In many western countries, allergic diseases in general are now the commonest type of chronic disease in children, and asthma is the commonest reason for emergency hospital admission and use of medications (Anderson 1997). In contrast, allergies are very rare in populations living subsistence lifestyles. Here the cause may be related to changes in exposure to infectious diseases, environmental and gut bacteria and parasitic worms during early life, although other hypotheses have also been posited. These ideas are considered in Chapter 7.

Mental health problems, particularly high rates of depression and stress, have also been associated with life in modern industrial societies. Furthermore, they contribute to other health problems, acting as risk factors for cardiovascular disease and probably for other diseases too. Changes in relationships with kin and in social organisation more generally have been blamed for rising rates of depression and stress. The case for this approach is examined in Chapter 8.

Chapter 9 offers a summary overview of what an evolutionary perspective teaches us about human vulnerability to western diseases. It goes on to look at projected trends in western diseases, especially in relation to the developing world. It is clear that western diseases are likely to become more and more important as populations become westernised and urbanised and as age profiles change so that the proportion of older people in the population increases. Here I also consider what an evolutionary perspective has to offer to the development of possible preventive strategies in the face of this growing problem.

I have not attempted to include a discussion of all western diseases, although I have covered most of the most important diseases. The obvious omission from the list given above is lung cancer, which is one of the biggest killers in most western societies and also elsewhere. I chose not to consider lung cancer in detail because it is so clear that smoking tobacco is its main risk factor. Instead, I have focused on those diseases for which an evolutionary perspective is particularly valuable.

Summary

There has been interest in the concept of western diseases and concern about the mismatch between human biology and the affluent western environment for nearly 40 years. This book sets out to apply and extend this evolutionary perspective, making use of our much increased understanding of evolutionary processes and of the biology and epidemiology of disease. I aim to show that western diseases, which constitute one of the major problems facing humans at the beginning of the twenty-first century, can be understood far more completely using this perspective.

2 An evolutionary history of human disease

In order to place western diseases in an evolutionary context it is necessary to consider the experience of human disease throughout human evolutionary history. To achieve this I adopt a framework drawn from the work of Boyden (1987) and Cohen (1989), illustrated in Table 2.1. This approach emphasises the need to understand the way of life, ecology and health experience of hunter–gatherer people, because such an understanding informs us about the context in which members of our species lived for so much of its evolutionary history. An examination of the enormous impact of agriculture and then of urban living on human health illustrates how changes in ways of life have had profound effects on disease experience in the past. As will become clear in later chapters, it is also increasingly apparent that an understanding of the evolutionary history of human exposure to infectious disease and nutritional pressures, which was profoundly affected by these innovations, is relevant to our understanding of western diseases. Finally, I consider the decline of infectious diseases and rise of non-communicable disease in the west, the so-called epidemiologic transition, and trends in the prevalence of western diseases over the nineteenth and twentieth centuries.

Human ecology and health in the Palaeolithic

Anthropologists have used various kinds of evidence to try to find out more about the ways of life of humans during this period, and to characterise their experiences of health and disease. The main sources of evidence are contemporary hunter–gatherers (also known as foragers) and relics of the Palaeolithic, such as fossils, prehistoric skeletal material, tools and other archaeological evidence. Neither type of evidence provides perfect information. There are few hunter–gatherer populations left, and these few have been profoundly affected by the economies and cultures of the people living around them, and are now generally limited to the harshest parts of their original terrain. As a result, they probably offer only a narrow view of how

9

Table 2.1. *Framework showing timescale of major cultural innovations that have affected human exposure to disease since the emergence of anatomically modern* Homo sapiens. *Generation time is estimated at 25 years*

Innovation	Years before 2000	Generations before 2000	Size of human communities
Emergence of anatomically modern *Homo sapiens* (living as hunter–gatherers)	100 000	4000	Scattered nomadic bands of 30–50
Development of agriculture	10 000	400	Relatively settled villages of <300
Development of cities and irrigated agriculture	5500	220	Few cities of 100 000; mostly villages of <300
Introduction of steam power	250	10	Some cities of 500 000; many cities of 100 000; many villages of 1000
Introduction of sanitary reforms	140	6	

Adapted from Mascie-Taylor (1993).

life might have been in the Palaeolithic. Relics of the Palaeolithic are limited. However, together these sources provide enough useful information to allow us to build up a picture of what life was like for modern *Homo sapiens*, and to a certain extent for earlier hominins, before agriculture.

During this time humans lived in band societies, foraging (mostly hunting and gathering) for food from wild resources. It is likely that a band consisted of around 30–50 people, and that bands were loosely coordinated into a network (Cohen 1989, p. 16). Hunter–gatherers hunt, fish and trap animals and gather or harvest wild fruits, roots, tubers, leaves and seeds. Scavenging from corpses of large animals may also have been important for the earliest *Homo sapiens*. Hunter–gatherers do not farm or raise domestic animals. Their population density is limited by their dependence on wild resources, which are generally more sparsely distributed than farmed foods. As a result they usually have to be mobile, moving on once the wild foods in an area are depleted either by exploitation or because of seasonal change (Cohen 1989).

Extrapolation from present-day hunter–gatherer societies also suggests that pre-agricultural human groups were characterised by egalitarianism; there is generally no dominance ranking and no evidence of institutionalised leadership amongst modern hunter–gatherers (Runciman 2005). The exact mechanisms maintaining such a social order, which contrasts with that of most non-human primates, are the subject of much debate (Erdal *et al.* 1994; Runciman 2005). In essence, though, the explanation appears to be that, in most hunter–gatherer societies, no individual can assume dominance because no individual can afford to risk alienating others and having them withdraw from exchange relations. This is because of the critical dependence, in the absence of storage or wealth (for example in the form of agricultural hold-ings), on the reciprocal exchange of everything valuable (Chisholm and Burbank 2001). However, men and women usually performed different tasks, classically men being the hunters of meat and women the gatherers of plant materials, carrying their infants and accompanied by young children (Hrdy 2000, p. 109).

Hunter–gatherer groups would have been too small and dispersed to sup-port many of the acute communicable diseases familiar to us today (Fiennes 1978). Acute respiratory diseases and many viral diseases, such as mumps and measles, require a pool of previously uninfected, and therefore suscep-tible, individuals, often children, to infect. They need to move to another host within the short infectious period and if they cannot do so they will die out in that population.

Pathogens with long periods of latency and low virulence, including a number of viruses, would have had more success (Cohen 1989; Van Blerkom 2003). For example, the herpes viruses, including *Herpes simplex* and the virus that causes chickenpox, can live within an infected individual for the rest of his or her life, infecting others when the disease becomes active again. Intestinal helminth (nematode worms and flatworms) infections also fit into this category. In fact, helminth parasites have been found in virtually every animal species examined and we can assume that such infections would have predated the divergence of the hominin lineage and continued to affect all subsequent hominins (Barrett *et al.* 1998). For example, it is likely that the pinworm or threadworm (*Enterobius*) was a parasite of hominins throughout their evolution (Kliks 1983). This point is important for current ideas about the causes of allergies (see Chapter 7). Pinworm is likely to have been particularly successful because a single individual host can continuously reinfect him or herself, especially when there is frequent hand to mouth contact, as in humans. This allows the transfer of eggs from the anus to the mouth. In addition, hunter–gatherers would have been susceptible to zoonotic diseases, that is diseases of other animals that intermittently affect humans and often cause severe illness. Examples of zoonoses are toxoplasmosis, rabies and also some helminths that are normally confined to non-human species.

The movement of hominid populations out of Africa, probably from around 50 000 years ago, would have exposed migrating bands to novel ecologies and associated pathogens. However, because of the low population densities and high mobility characteristic of hunter–gatherers, it is likely that a similar kind of disease ecology would have emerged in each new environment (Barrett *et al.* 1998).

In most contemporary hunter–gatherer groups a great diversity of food sources is used (O'Dea 1991) and diseases associated with deficiency of particular nutrients are rare, as are undernutrition and overnutrition. A detailed consideration of hunter–gatherer diets is provided in Chapter 3.

Despite this generally positive picture of health, data from contemporary hunter–gatherer groups suggests that it is likely that infant and childhood mortality was fairly high, and that childhood growth was slow in comparison with modern reference standards, which are based on US children (Stinson 2002). Estimates of infant mortality in contemporary hunter–gatherer groups range from around 150 to 250 deaths per thousand (Cohen 1989, p. 100), a figure which is very high in relation to current rates in affluent countries (around 10–12 per thousand) but fairly similar to rates found historically in Europe and currently in many poor countries. Biesele and Howell (1981) suggested that about half the nomadic !Kung of the Kalahari Desert, probably the best-known hunter–gatherer group, died before the age of 15. Cohen (1989, p. 129) suggested that the average life expectancy at birth for hunter–gatherers was about 25–30 years. This figure is similar to, or slightly higher than, estimates for much of Europe until the early nineteenth century. Those who survived to adulthood had a good chance of reaching old age, with a few surviving into their eighties.

So, how did hunter–gatherers die during the Palaeolithic? Deaths in hunter–gatherer societies today are due mainly to accidents (often involving fire), trauma and violence, and infectious disease (Cohen 1989, p. 92) and it seems likely that this would also have been the case in the past. Disablement due to injury or illness would have been much more likely to lead to death than in societies with developed health care systems, and the need for mobility would have threatened those who were not fully physically active (Boyden 1987, p. 76). An intensive study of the Ache in Paraguay found that the main causes of death when they lived as forest-dwelling and largely isolated hunter–gatherers were death at the hands of another human (including infanticide), accident (snakebite was the most common cause of accidental death), fever and uncategorised illness (Hill and Hurtado 1996).

In summary, the disease profile of hunter–gatherer populations appears quite positive. Epidemic infectious disease would have been unlikely and it is also clear that cardiovascular disease, type 2 diabetes, reproductive cancers, other cancers associated with obesity, and allergies would have been absent or virtually absent in all these groups (Cohen 1989; Hurtado *et al.* 1996). Nevertheless, we should not overstate the picture of a disease-free life; infections are likely to have been a significant cause of illness and to have imposed some selective pressures at this stage of human evolution (Van Blerkom 2003).

The impact of early agriculture

From around 14 000 years ago, the earth experienced a period of large climatic fluctuations, changes that had a huge impact on plant and animal ecology (McMichael 2001, p. 136). In response, some human groups began to domesticate plants and animals, starting about 10 000 years ago (Cohen 1989; Harris and Hillman 1989). This happened first in the Old World and then, several thousand years later, in the New World. The early farmers necessarily became more sedentary and settlements started to appear. This change is known as the Neolithic revolution, although its pace was slow.

There was a huge shift in disease patterns as a result of the more settled nature and larger density of human populations and the domestication of some animals. Sedentism encourages the transmission of diseases that benefit from accumulation of waste and close proximity to domestic animals. These include vector-borne diseases, particularly those carried by insects and rats, which are attracted to human shelters and waste. For example, yellow fever is transmitted by a mosquito that almost always lives in water stored around human dwellings (Cohen 1989). Humans also acquired a large number of helminth parasites from domestic animals at this stage, including *Ascaris* (roundworm) and *Trichuris* (whipworm), now the most common soil-transmitted helminths in humans, from pigs (Kliks 1983). There is good evidence that helminth infections in early agriculturalists were considerably greater than in hunter–gatherers inhabiting a similar area (Reinhard 1988).

There are several pathogens with complex or extended life cycles outside human hosts that are disrupted in nomadic people, but can be successfully completed within a sedentary agricultural population (Mascie-Taylor 1993). Hookworm, a common and disabling infection in many poorer societies today, is an example of such a parasite. Schistosomiasis or bilharzia is also an important disease today in many parts of the world. Its continued transmission is only possible if humans use a source of fresh water on a regular basis, as sedentary populations must. Malaria is another example, thought to have become prevalent only with the introduction of agriculture partly because agriculture provided conditions, such as pools of standing water, allowing the spread of the mosquito populations that transmit the disease (Relethford 1994; Pearce-Duvet 2006).

Consumption of meat probably declined with the beginnings of agriculture, with important implications for dietary quality (Larsen 2003). Animal foods are excellent sources of protein for humans, since they have a balance of amino acids suited to human needs, and they are also good sources of a number of vitamins and minerals (Larsen 2003). Animal fat is a particularly rich source of calories. Dependence on a relatively small number of domesticated plants would also have impaired dietary quality by restricting the variety of micronutrient intake (Eaton *et al.* 2002). Furthermore, the cereals that are commonly staples in the diets of agriculturalists all have nutritional shortcomings. For example, wheat-rich diets, found in temperate Asia and Europe, are associated with zinc deficiency, and dependence on rice can cause beriberi (thiamine deficiency) in certain circumstances. Dependence on sorghum, which is an important staple grain in north Africa, can result in pellagra (niacin deficiency). Some, but not all, small-scale farming populations dealt with this successfully by developing techniques to overcome their specific problems; for example, the consumption of beans with maize by Native Americans compensated for the deficiencies of maize, which is poor in some amino acids, niacin and iron.

Because of an increase in population size and their dependence on a limited number of crops, agriculturalists are vulnerable to crop failures and seasonal fluctuations. This means that they probably experienced as least as much, and probably more, shortages in food supply than did hunter–gatherers (Prentice *et al.* 2005; Benyshek and Watson 2006). This point is important in relation to ideas about 'thrifty genotypes' and susceptibility to type 2 diabetes, as outlined in Chapter 4. It is not surprising then that general malnutrition is more common amongst agriculturalists than amongst hunter–gatherers (Barrett *et al.* 1998). In addition, the poor nutrition experienced by early agricultural populations would have made them more vulnerable to infectious diseases.

Iron-deficiency anaemia, a disease that may have no symptoms, but which often causes tiredness, paleness and weakness, is thought to have become a cause of ill-health for humans with the adoption of agriculture. Evidence that iron-deficiency anaemia became more common when farming was introduced is provided by the skeletal pathology porotic hyperostosis in the remains of early agriculturalists. This is a condition in which the normally smooth, dense outer compact bone of the skull and orbit is pitted by small holes (Stuart-Macadam 1998). Contemporary hunter–gatherers who still follow traditional lifestyles do not suffer from iron-deficiency anaemia. One suggestion is that the cause of the disease in early agriculturalists was the increased load of infectious diseases associated with farming, particularly repeated diarrhoeal infections, which cause bleeding and prevent proper nutrient absorption in the gut (Kent 1986). Others suggest that the main cause was dependence on staple foods that are low in iron (such as maize), or that inhibit the absorption of iron (such as maize, millet and sorghum). Probably a combination of these factors was responsible (Holland and O'Brien 1997). This debate illustrates the importance of both infectious disease and nutritional problems for agriculturalists.

New infectious diseases imposed selection pressures on early farmers, and, where these pressures were severe enough, it is likely that genes that conferred resistance spread in affected populations. The classic example is the spread of genes that protect people against malaria (Smith 1993). The selective pressure exerted by malaria is strong since it kills large numbers of people, particularly infants and young children. Anybody who carries genes that allow them to survive the disease is particularly likely to pass those genes on so that they spread through the affected population, despite having seriously adverse effects if two anti-malarial genes are inherited together. Changes in diet associated with taking up agriculture are also likely to have imposed selective pressures. Thus, while most humans cannot digest lactose as adults and cannot therefore benefit from drinking milk, many populations that have relied on milk consumption for thousands of years have evolved the capacity to digest lactose by retaining production of the enzyme lactase in adult life, and obtain significant nutrition from milk (Durham 1991, p. 237). Europeans, Tanzanians, Kenyans and Sudanese all have long histories of milk consumption and a large number of people from these populations produce lactase in adult life. These populations carry different variants of the lactase gene, but all are thought to have arisen because of the strong selective pressure resulting from animal domestication and adult milk consumption (Tishkoff *et al.* 2007).

Dental health also declined with the shift to farming, as evidenced by skeletal material (Larsen 2002). For example, dental caries is more common in teeth from early agriculturalists than from the hunter–gatherers who preceded them (Larsen 1995). The increase in dental caries is thought to have been caused by the increased consumption of plant carbohydrates, which are fermented by bacteria, producing acids that cause tooth decay. This change appears to be most marked in New World populations, where agriculturalists relied heavily on maize, and is less clear-cut in other parts of the world (Eshed *et al.* 2006). Tooth loss as a result of gum disease was also more common in early agriculturalists than in hunter–gatherers. Another marker of poor health in early agriculturalists is the fact that they generally have more enamel hypoplasias than hunter–gatherers (Larsen 1995). These horizontal grooves or pits in the enamel provide a good record of nutritional and other stresses, including infections, experienced during childhood.

Childhood growth is a very sensitive indicator of a population's nutritional status. We have already seen that childhood growth is and was slow amongst hunter–gatherers, compared with modern US standards. In early agriculturalists, growth rates seem to have been even slower. For example, in the Illinois Valley in the USA, growth rates were slower in early agriculturalists than in hunter–gatherers (Cook 1984). Children who grow slowly are more likely to be short as adults and adult height was shorter in early agriculturalists than in hunter–gatherers from the same area in many parts of the world (Larsen 1995).

Skeletal studies also suggest a lower average life expectancy in early farmers compared with hunter–gatherers (Larsen 1995). However, this is a complex area and the available evidence, all of which is indirect, can be interpreted in a number of ways (Bentley *et al.* 2001). Thus it is difficult to assess changes in mortality associated with the Neolithic revolution.

In summary, the development of farming radically changed the disease experience of humans. By providing pathogens with new niches, it allowed many important infectious diseases to emerge. In addition, diet-related diseases became important for the first time, as people moved away from a reliance on wild food sources. There is a general consensus that the health of early farmers was worse than the health of hunter–gatherers, giving us our first example of how a large shift in lifestyle can affect health.

Urban societies

The first cities developed in Mesopotamia and Egypt around 5000 years ago, based on the development of irrigation and associated increases in agricultural yields (Carter 1999). In cities, population density was greater than previously, and, in contrast to earlier types of settlement, most of the population were not farmers, so that the urban population was dependent on the food production of those living in rural areas. People living in cities were also stratified according to wealth and power, leading, probably for the first time, to different experiences of health and disease in different sectors of the population (Boyden 1987, p. 128).

Increased population density creates even more possibilities for transmission of infectious disease. Many of the infectious diseases now most familiar require a large, dense, population, of the type only found in cities. This is particularly the case for those viruses which, if they do not kill their host, give rise to long-lasting immunity (Boyden 1987). Examples are the viruses that cause measles, mumps, whooping cough and poliomyelitis (Baker 1988). Diseases spread by the faecal–oral route, such as cholera and typhoid, have the opportunity to cause epidemics in crowded and unsanitary living conditions, where water supplies are often contaminated. Reliance on food produced in the countryside also creates vulnerability. For example, in pre-industrial Europe, there was serious malnutrition when harvests failed and food was scarce (Wrigley 1969). Omran (1971) cites figures from Graunt's seventeenth-century study of London's *Bills of Mortality*, which suggest that nearly three-quarters of all deaths in London at that time were attributed to infectious diseases, malnutrition and 'maternity complications', while cardiovascular disease and cancer were responsible for less than 6% of deaths.

Long-distance travel and trade between cities aided in the transmission of infectious disease. For example, trade between city states allowed the initial spread of bubonic plague from Asia to Europe. Crowded and unsanitary mediaeval cities such as London provided ideal conditions for the transmission of plague – the wattle and daub houses offered a good habitat (including access to stored grain) for the black rats that carry the fleas that transmit plague (McElroy and Townsend 2004, p. 159). During the worst outbreak of the Black Death in Europe between 1347 and 1351 (probably, but by no means certainly, caused by bubonic plague) between 25% and 75% of the population, at least 25 million people, are believed to have died from the disease (Scott and Duncan 2001, p. 87). From the seventeenth century onwards, the exploration and colonisation of the Americas by European powers introduced previously unknown diseases such as measles and smallpox, to devastating effect.

Cities also have the capacity to pollute the air in ways that can affect health. At least as far back as the thirteenth century, English royalty escaped polluted air by extended visits to the countryside (Strassmann and Dunbar 1999). In London, the great smog of 1952 was responsible for around 4000 deaths, mostly in the elderly or in those with heart or respiratory system complaints (Elsom 1992, pp. 242–243). In the worst affected American cities today particulate pollution reduces life expectancy as a result of the aggravation of cardiopulmonary disease (Strassmann and Dunbar 1999). Associations between air pollution and asthma are considered on pp. 124–125.

It is possible to make an assessment of overall health from the skeletal remains of those who lived in cities and to compare them with rural agriculturalists and hunter–gatherers. Thus Steckel *et al.* (2002) proposed the construction of a health index from skeletal evidence of stature, signs of iron-deficiency anaemia, enamel hypoplasia, dental health, degenerative joint disease, trauma and skeletal infections. They found that residence in a town or urban area was an important determinant of a poorer health index, with the worst level of health seen in those living in these large settled communities (Steckel and Rose 2002a).

Mortality rates were high and life expectancy was low in cities prior to the industrial revolution (Wrigley 1969, p. 131). For example, Storey (1985) estimated mortality rates from a skeletal sample from Teotihuacan in central Mexico, the earliest urban society in Mesoamerica, existing between 150 BC and AD 750. Even though the New World is thought to have been relatively free from epidemics of infectious disease prior to contact with Europeans, city-dwellers there are thought to have suffered from malnutrition and endemic disease. Life expectancy at birth in the sample from Teotihuacan is estimated at 17 years, while life expectancy at age 15 was 35. Almost a third of the birth cohort had died by the age of 1 year and only 46% remained by age 15. This picture of high infant and juvenile mortality, with a fairly low adult life expectancy is similar to that in both imperial Rome and seventeenth-century London (Storey 1985). Thus urban mortality and life expectancy were probably worse than those experienced by hunter–gatherers.

Early city life introduced more new diseases for humans. Life was short and hard for most people, and there was a high risk of infectious disease and malnutrition.

The epidemiologic transition

Mortality from infectious disease started to fall at about the time of the industrial revolution in Europe and North America. First, epidemics of diseases such as typhoid, cholera, measles, diphtheria and whooping cough started to decline (Seale 2000), as did famine, reducing episodes of 'crisis' mortality (Anderson 1988). Then, between about 1850 and 1950, there was a gradual decline of mortality from non-epidemic infectious diseases such as pneumonia, bronchitis, diarrhoeal diseases and tuberculosis (Barrett *et al.* 1998; Seale 2000). Reductions in child and infant mortality were particularly marked (Schofield and Reher 1991). The factors responsible for this reduction in infectious disease mortality include improved nutrition associated with improved agricultural yields and distribution networks, pasteurisation, public hygiene measures, primary health care and, at a later stage, some biomedical advances, such as the vaccination campaign against smallpox (Schofield and Reher 1991; Barrett *et al.* 1998). The relative importance of these factors, however, continues to be debated.

As a consequence of the decline in mortality from infectious diseases and the associated improvement in survival through infancy and childhood, life expectancy at birth increased dramatically during this period in the western world, by around three decades between about 1850 and 1950 (Fig. 2.1) (Seale 2000; Wilmoth 2000). This was probably the first time in human experience that life expectancy at birth rose above 20–30 years.

During this period, non-communicable diseases of middle and later adulthood started to replace infectious diseases as the major causes of morbidity and mortality. This shift was described by Omran (1971) as the final stage of the epidemiological transition from the predominance of infectious diseases and famine as causes of mortality to the predominance of 'degenerative and man-made diseases'. Omran (1971) noted that, in England and Wales, heart disease and cancer together accounted for gradually more deaths

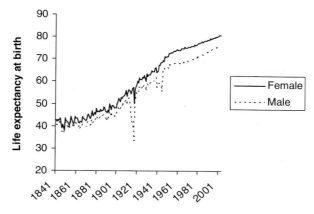

Fig. 2.1 Life expectancy at birth for England and Wales from 1841 to 2003. Data from the Human Mortality Database, University of California, Berkeley (USA) and the Max Planck Institute for Demographic Research (Germany). Available at www.mortality.org (data downloaded on 12th January 2007).

until about 1920 and then increased in relative importance more quickly into the second half of the twentieth century. In 1921 cardiovascular disease and cancer combined caused approximately the same number of deaths in Britain as infectious diseases, but by 1931 the chronic diseases had become the major causes of death in men and women (Davey Smith and Marmot 1991).

By the end of the twentieth century, fewer people were dying of infectious disease and more of non-communicable diseases in the poorer parts of the world as well. The World Health Report of 1999 reported that non-communicable diseases were responsible for 59% of total global mortality (World Health Organization 1999). The decline of mortality from infectious diseases and rise of mortality from non-communicable diseases is illustrated for a middle-income country, Chile, in Fig. 2.2. In contrast to the transition experienced in Europe and the United States, which largely preceded modern biomedicine, treatments such as oral rehydration therapy, immunisations and antibiotics played an important role in the initial successes in mortality reduction in the poorer nations (Barrett *et al.* 1998). Reflecting these changes, average life expectancy at birth worldwide rose from 48 years in 1955 to 65 years in 1995 (Seale 2000).

However, the view that the epidemiologic transition, as described by Omran, is now being experienced in poorer parts of the world has been heavily criticised, in large part because infectious diseases and diseases of undernutrition have failed to disappear. For example, infectious diseases, maternal and perinatal causes and nutritional deficiencies still account for the

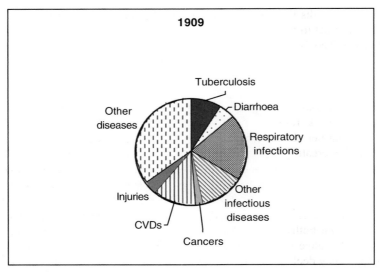

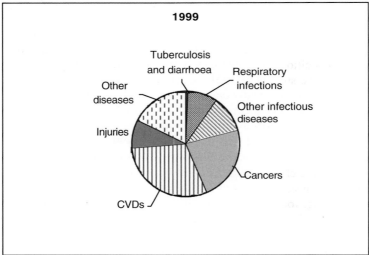

Fig. 2.2 Change in the profile of causes of death in Chile between 1909 and 1999. In 1909 47% of deaths were caused by infectious disease and 15% by cancer and cardiovascular disease. In 1999 21% of deaths were caused by infectious disease and 53% by cancer and cardiovascular disease. CVDs = cardiovascular diseases. Reproduced from McMichael (2001) with permission from Cambridge University Press.

majority of deaths in India and sub-Saharan Africa (Heuveline *et al.* 2002). The main threat to the epidemiologic transition in many parts of the world is HIV/AIDS. This is especially the case for sub-Saharan Africa, where AIDS is having a huge impact on life expectancy. For example, in the nine African countries with adult HIV prevalence of 10% or more, life expectancy by 2010–15 will be 47 years on average, compared to the 64 years that it would have been without AIDS (Seale 2000). In addition, other infectious diseases, often known as re-emerging infections, are regaining importance. For example, warmer climates and seas have led to increased coastal blooms of algae, which create favourable environments for the proliferation of cholera (Patz *et al.* 1996). Richer countries have also been affected by a resurgence of some infections. The most dramatic example is tuberculosis, including multi-drug resistant tuberculosis (Gandy and Zumla 2002).

The situation in Eastern Europe is different again, with rising levels of mortality from both non-communicable and infectious disease presenting a different challenge to the model of the epidemiologic transition. In Russia, life expectancy declined between 1987 and 1994, partly as a consequence of an increase in mortality from cardiovascular disease, but also because of increases in mortality from infectious diseases and accidents and violence (Leon *et al.* 1997). These declines in life expectancy have been caused partly by deterioration in public health services as well as by worsening material and social conditions (Seale 2000). Thus it is also difficult to fit these countries into the simple model of epidemiologic transition.

The rise of western diseases

If people do not die of infectious disease, and so live long enough to become susceptible to the chronic degenerative diseases of old age, they will die of diseases of old age. It is useful to ask, therefore, whether there was a real, age-specific (e.g. for a 60-year-old) increase in the risk of dying from non-communicable disease associated with the epidemiologic transition. The answer appears to be that there was not, and that there was, in fact, a decrease in the age-specific risk of dying from non-communicable disease as well as from infectious disease between the nineteenth and twentieth centuries (Preston 1976; Gage 2005). This phenomenon has been investigated in some detail using data from the United States (see pp. 45–46) and the general conclusion is that the decline in manual work and in infectious disease over this time period contributed to decreased age-specific rates of some non-communicable diseases.

Thus the most important factor in the rise of non-communicable disease has been the increasing age of populations. Nevertheless, it seems clear that, over the first half of the twentieth century, there was an increase in age-specific mortality from coronary heart disease and from some cancers, the most important of which at this time was lung cancer (as a result of increased smoking) (Davey Smith and Marmot 1991; Gage 2005). That is, the risk of contracting these diseases at a given age over this period really did increase. Subsequently, rates of both coronary heart disease and lung cancer went into decline in the west, as people heeded health messages about smoking in particular, but they are now increasing elsewhere. The other classic western diseases have become more common in most places over the second half of the twentieth century. For example, age-specific rates of breast cancer have been increasing for many years in affluent countries such as those in Europe, with more new cases over time, even after taking into account the fact that the population has been ageing (Botha *et al.* 2003). Over the second half of the twentieth century there has also been a large increase in the number of new cases of type 2 diabetes per year in the United States (Fox *et al.* 2006). Rates of clinical depression and allergy have increased over this period in western countries and elsewhere (Klerman and Weissman 1989; Masoli *et al.* 2004a). Most recently, there has been considerable concern about rapidly rising levels of obesity, especially in the young, in many parts of the world.

Summary

In the New World, Steckel and Rose's skeletal health index declined 'as Western Hemisphere societies evolved from simple to complex and hierarchical ... In general, the healthiest populations were hunter–gatherers, and the least healthy lived in large settlements supported by systematic agriculture' (Steckel and Rose 2002b, p. 587). This is the story of human health at its simplest, reflecting the increasing impact of infectious disease and dietary deficiencies, up until the Industrial Revolution. From that point onwards, an increased pace of technological change in the west ushered in a reduction in the importance of infectious disease and a rise in the importance of non-communicable disease. This epidemiologic or health transition has not been repeated in a simple way in other parts of the world, but non-communicable diseases have also increased in importance in Eastern Europe and the poor and middle income countries, so that they are currently the biggest cause of death in the world. We start the twenty-first century with huge concerns about the increasing rates of non-communicable western diseases around the world.

3 Obesity, type 2 diabetes and cardiovascular disease

Obesity is common and increasing in prevalence in Europe, North America, Australia and Latin America, particularly among the young (Seidell 1995; Martorell *et al.* 1998; Midthjell *et al.* 1999; Catanese *et al.* 2001; Tremblay *et al.* 2002; Thorburn 2005). While obesity emerged first in western populations, it is rapidly spreading to poorer countries (Yoon *et al.* 2006). Considerable concern has been expressed in the medical literature, as well as more widely, about the obesity 'epidemic' and its associated health consequences. In this chapter I outline the evolutionary context for the emergence of obesity as a major health problem, comparing what we know about diet and energy expenditure, the main determinants of obesity levels, in hunter–gatherer societies and in industrialised countries today. I also examine the closely related diseases type 2 diabetes and cardiovascular disease. Some populations seem to be particularly vulnerable to the development of obesity and related diseases, an issue explored in Chapter 4.

Diet, physical activity and body composition in humans before agriculture

The first members of the taxonomic group to which apes and humans belong (hominoids) lived 30 million years ago and subsisted largely on fruit, with some vegetable foods and meat. Meat began to assume more importance in the hominin line from around 7 million years ago, particularly with the advent of stone tools about 2 million years ago (Richards 2002). The increasingly gracile crania of hominins are regarded by many as evidence of this dietary trend because they are interpreted as being associated with a reduction in chewing as a consequence of eating more meat and less fibrous plant material. Stone tools and finds of animal remains at 'home bases' of *Homo* species also provide evidence of meat consumption. Evidence from the bone chemistry of anatomically modern *Homo sapiens*, living in Europe between 30 000 and 10 000 years ago, suggests that their diets contained high

23

levels of animal protein (Richards 2002). Thus early humans had a much higher quality, more nutrient-dense, diet than other primates (Leonard 2000). Adaptation to this diet is reflected in our gut morphology, which in some respects is more like that of a carnivore than that of the more characteristic primate folivore or frugivore (Leonard 2000). As we saw in Chapter 1, this dietary adaptation is thought to be associated with the evolution of increased brain size in the hominin line, requiring a more easily digestible and energy-rich diet (Aiello and Wheeler 1995).

The ecology of food acquisition has been central to studies of hunter–gatherers. Consequently, we know a great deal about the different ways in which extant hunter–gatherers acquire their food, and about the composition of their diets and their levels of energy expenditure. These studies provide the best information available to help us estimate the nutritional ecology of our ancestors during the Palaeolithic. There is a great deal of variation in diet between groups, but it is possible to make some generalisations which give insights about the reasons for the current rise in obesity and associated health problems.

Diet

O'Dea (1991) provides a picture of the traditional lifestyles of Australian Aborigines living as hunter–gatherers in remote parts of Australia in the twentieth century, which serves as a useful illustration against which to consider more general trends. Australian Aborigines were omnivorous and the composition of their diet varied greatly according to their geographical location and by season. They relied on an intimate knowledge of the land, of sources of fresh water, and of the plants and animals in their territory. Women gathered plant foods, honey, eggs, small mammals, reptiles, fish, shellfish, crustaceans and insects. Men provided less regular but highly valued food supplies such as kangaroos, large birds, reptiles and fish. Usually, there was one main meal per day, eaten when people returned to camp in the late afternoon. However, throughout the day, while hunting and gathering, people would snack on foods such as grubs, fruits, gum, honey ants and honey (Fig. 3.1). Everything edible on an animal carcass was eaten. Cooking processes for large game were often ritualised, for example, turtle was cooked directly on coals, with hot stones placed inside the abdomen. Food was usually eaten more or less immediately, but in some circumstances foods were dried or processed in some other way for storage. The most valued components of the diet were energy-dense foods such as animal fat, organ meats, fatty insects and honey. These foods were available either only seasonally or in small quantities. Traditionally, Australian Aborigines lived in bands based on extended family groups of around 20–30 people,

Fig. 3.1. Aboriginal boys of the Maragunggu group near Makanba Creek, Northern Territory, Australia with foraged wirlgngu ('red apples', *Syzygium suborbicularis*) in 1979. Traditionally, this fruit is eaten raw and is prepared in a variety of ways for medicinal use (Brock 1993). Photograph courtesy of Bob Layton.

but they were able to gather in larger groups for ceremonies if food was available to support these larger numbers. This was possible in southern Queensland, for example, when Bunya pine nuts were ripe.

Other hunter–gatherer groups share some of these characteristics of subsistence, particularly the gender division of labour, but clearly the resources used differ considerably in different environments around the world. Similarly, there must have been substantial variability in dietary composition among early human forager groups (Jenike 2001). Nevertheless, early human diets shared characteristics that distinguish them from current diets.

Eaton *et al.* (1999) estimate that daily caloric intake was perhaps greater than that of the average US citizen today, but Cohen (1989, p. 95) notes 'hunter–gatherer diets that are otherwise well balanced can be poor in calories because wild game is lean, wild vegetable foods often contain a very high proportion of bulk or roughage, and concentrated packages of calories – such as honey – are rare in the wild'. Even at the levels of intake estimated by Eaton *et al.* (1999), it is not clear that there was always enough food available to meet the energetic needs of all individuals, since hunter–gatherers are much more physically active than westerners today. However, it is likely that caloric availability was adequate in comparison to standards in the poorest populations today.

There is some debate about the composition of the early human diet. For some time, the most common estimate has been that, on average, hunter–gatherers obtain 35% of their energy intake from meat and 65% from plant food (Eaton *et al.* 1988a, p. 80). However, a recent re-analysis of data on contemporary hunter–gatherer diets suggests that there has been too great an emphasis on the role of plant foods and that animal-derived products contribute around 65% of energy (Cordain *et al.* 2000). Cordain *et al.* argue that, whenever and wherever it was possible, hunter–gatherers would have chosen to consume large amounts of animal food because it provides greater energy returns than plant food in relation to the energy required for its acquisition. They suggest that large prey species would have been preferred because they provide the best energy returns per kill and because they have an increased percentage of body fat. The higher proportion of body fat provides extra energy and helps protect against potential health problems associated with high levels of protein consumption (Cordain *et al.* 2000). These can arise because there appears to be a limit to the rate at which the gastrointestinal tract can absorb amino acids from dietary proteins and to the liver's capacity to process proteins and excrete excess nitrogen (Bilsborough and Mann 2006).

Cordain *et al.*'s analysis suggests that, for the majority of hunter–gatherer societies, dietary fat intake contributes between 36 and 43% of total energy, a level quite similar to that currently seen in the USA. They also suggest that levels of dietary protein would have been higher than those seen today, perhaps 19–35% of dietary energy compared to 16% in western populations. Estimates of carbohydrate consumption vary from Eaton *et al.*'s (1999) suggestion of 45–50% of total energy intake, a percentage similar to that seen in affluent nations today, to Cordain's (2000) more recent estimate of 22–40% of total energy. However, some groups, such as the !Kung and the Hadza, rely more heavily on plant foods, while others, living in the Arctic, rely totally on animal foods, leading to great variation in macronutrient composition.

It is likely that the most pertinent characteristics of the hunter–gatherer diet are not so much overall macronutrient composition, but the reliance on wild plant and animal foods. Both wild plant and animal foods are generally very different in composition to their domesticated and processed counterparts.

Fat from wild animals and fish in the diet of hunter–gatherers tends to contain more polyunsaturated and monounsaturated fatty acids, and to have a higher ratio of omega-3 to omega-6 polyunsaturated fatty acids than the contemporary western diet (Sinclair and O'Dea 1993; Cordain *et al.* 2002). This fatty acid composition appears to have beneficial effects on serum lipid levels and type 2 diabetes risk (Eaton *et al.* 1999; Mann 2002). Neither dairy food, which is high in saturated fats, nor *trans* fatty acids, a particularly harmful component of processed foods produced by the hydrogenation of fats, would have been present in the diets of hunter–gatherers (Eaton *et al.* 1999). Thus it is likely that differences in the types of fats consumed are more important than differences in total fat consumption. Further evidence for this suggestion comes from findings that serum lipid profiles of Greenland Inuit are healthier than those of Europeans, despite the fact that Inuit eat only animal products, many of which are high in fat (Cordain *et al.* 2002).

The sources and types of carbohydrates consumed by hunter–gatherers are also markedly different from all other groups, with most of their carbohydrates coming from vegetables and fruit, rather than from grains, which became the main human source of carbohydrates with the beginning of agriculture. Although many groups used wild honey as a seasonal delicacy, it was obtained only with considerable effort, including confrontation with bees (Eaton *et al.* 1999). Thus simple sugars would have been eaten in very low quantities.

Typically, levels of fibre are high in the contemporary hunter–gatherer diet and archaeological evidence derived from fossilised faeces also suggest that levels were high in the Palaeolithic diet (Eaton *et al.* 1999). Food was eaten raw or was processed by pounding, scraping, roasting or baking, none of which removes fibre. High levels of consumption of fruit and vegetables are known to protect against cardiovascular disease in a western context (He *et al.* 2006; Kaliora *et al.* 2006). In addition to fibre, other components of fruit and vegetables thought to be beneficial in relation to cardiovascular disease are potassium, folate and antioxidants, including vitamin C (He *et al.* 2006).

Hunter–gatherer diets vary seasonally with plant and animal productivity. In contemporary hunter–gatherer groups for which good information on seasonality of the diet is available, protein intake generally remains adequate all year round, but intakes of total energy and fats and carbohydrates fall during seasons when there is low availability of plant foods and wild animals are particularly lean (Jenike 2001). Archaeological data have also been interpreted as suggesting seasonal food shortages. As noted below, the occurrence of seasonal and other occasional shortages in energy, fat and carbohydrate consumption among most hunter–gatherer societies, may be linked to the capacity of humans to store and mobilise large quantities of body fat (Brown and Konner 1987).

We can also consider the human preference for high fat, high sugar, salty foods in the light of our knowledge about diet before the adoption of agriculture. These preferences probably evolved because they stimulate the consumption of foods high in energy and nutritional quality (for example, meat is salty) (Boyden 1987; McMichael 2001, p. 264). In fact, these tastes probably predate the origin of hominins, since many mammalian species, including modern apes, also show a preference for fats, for example, in seeking out lipid-rich nuts and fatty insects (Pond 1998, p. 224). It may also be that the frugivorous diet of our ancestors was associated with the enjoyment of alcohol. Dudley (2000) argues that a liking for alcohol may have encouraged the seeking of ripe fruit, which contains significant amounts of ethanol, as well as having a high nutritional value. All these preferences would have been adaptive in environments in which levels of total energy, fats and carbohydrates were constrained.

Energy expenditure

O'Dea (1991) provides a brief picture of the types of physical activity in which Australian Aborigines, living as traditional hunter–gatherers, engage. She states that 'food procurement and preparation for Aboriginal hunter–gatherers were energy-intensive processes that could involve sustained physical activity: walking long distances, digging in rocky ground for tubers deep below the surface, digging for reptiles, eggs, honey ants and witchetty grubs, copping with a stone axe (for honey, grubs, etc.), winnowing and grinding of seeds, digging pits for cooking large animals, and gathering wood for fires (for cooking and warmth)'.

Amongst the forest-living Ache foragers in Paraguay, men hunt for about 7 hours per day, often running in pursuit of mammalian game, which are killed with a bow and arrow (a post-Palaeolithic innovation). They also collect honey when available. Women extract fibre from palm trees, and gather fruits and insect larvae, spending about 2 hours per day collecting food, in addition to their childcare activities. Women also carry the family's possessions, children, and pets in woven palm carrying baskets, spending an additional 2 hours per day on the almost daily camp move (Hill and Hurtado 1996, p. 65). In the 1970s !Kung women, living in the Kalahari desert, walked from 2 to 12 miles, 2 to 3 days a week, returning home with 7–15 kg of plant food (Lee 1979, p. 310). They also carried infants and sometimes young children. As with Australian Aborigines, gathering food was often demanding, requiring extensive digging through hard earth for tubers and roots. In addition, the ! Kung made day trips, sometimes once or twice a week, to visit friends and relatives in camps that were typically 3 to 10 miles away. In general, adult hunter–gatherers require stamina for food procurement, and moving camp often demands considerable physical exertion (Eaton *et al.* 1988b, p. 178). Children in hunter–gatherer societies are similarly active, engaging in vigorous physical play, unless they are accompanying their parents on hunting or gathering expeditions, visits or moving camp (Eaton *et al.* 1988b, p. 178).

A commonly used measure of physical activity, which allows comparisons of physical activity levels to be made between populations of different body size and composition, is the PAL (physical activity level). PALs express the energy costs of daily activities relative to basal metabolic rate, the energy required to maintain bodily functions, for a particular population. PAL can be calculated in different ways, most commonly by observing the time spent in particular activities and calculating the likely energy expended, but also by heart-rate monitoring. Although useful, these methods have their limitations, so that the figures obtained must be treated with some caution (Schutz *et al.* 2001).

As with diet, there is considerable variation in energy expenditure between hunter–gatherer groups and we should bear in mind that most contemporary hunter–gatherers live in marginalised environments where it is particularly hard to make a living. Sackett (1996, cited in Jenike 2001) calculated a composite hunter–gatherer PAL from published energy expenditure and time allocation information (that is, the amount of time spent on different activities) for a number of hunter–gatherer groups. These were 1.78 for men and 1.72 for women. According to the FAO/WHO/UNU (FAO/WHO/UNU Expert Consultation 1985) guideline figures, these indicate moderate levels of physical activity. Figures calculated in the same way for horticulturalists (1.87 for men and 1.79 for women, also moderate) and agriculturalists (2.28 for men and 2.31 for women, falling into the heavy activity category), suggest that Sahlins (1972) was correct in his assertion that hunter-gatherers do not have to work as hard for a living as those who cultivate the land. However, all these groups are considerably more physically active than people with sedentary jobs in industrialised economies. Shetty *et al.* (1996) suggest that a PAL of 1.55–1.65 (a light activity level) represents the average for 'the so-called sedentary lifestyle' in affluent societies. Measures of physical fitness are also much higher for hunter–gatherers than for typical North Americans (Eaton and Eaton 1999b). For example, fitness tests conducted in 1969 and 1970 in an Inuit population relying substantially on hunting, trapping and fishing showed that the active lifestyle was associated with higher levels of aerobic power and leg muscle strength than seen in white North Americans of the same age (Rode and Shephard 1971).

Physical activity is not the only source of energy expenditure for humans. For example, we also need energy to maintain homeostasis, including a stable body temperature. Hunter–gatherers need to expend much more energy in thermoregulation than contemporary humans, particularly those living in affluent societies who spend most of their time in permanent structures with regulated indoor temperatures. Pregnancy and lactation also require considerable energy, perhaps 80 000 kcal for one pregnancy (Brown and Konner 1987) and more to breastfeed the infant for the prolonged period typical of hunter–gatherers. Since most women in hunter–gatherer groups spend most of their reproductive lives either pregnant or breastfeeding (see Chapter 6) this is a substantial energetic cost for women.

Body fat and pathologies associated with obesity

All humans, including hunter–gatherers, have a proportionally greater body fat than other species occupying the tropical savannah environment where the genus *Homo* evolved (Wells 2006). Some other mammals do lay down a lot of fat, but in very different circumstances from humans, for example, marine mammals require fat for insulation, while other mammals require a supply of energy to maintain them through hibernation. In humans, an important function of fat is as an energy store. In such circumstances, evolutionary theory predicts that fat deposition is favoured by selection when reproductive success benefits more from buffering fluctuations in energy supply than from converting all energy intake directly into larger size or additional offspring (Wells 2006). In humans the need to buffer fluctuations in energy supply derives from a number of pressures.

One function of fat in many mammals is to sustain animals through seasonal declines in food availability. Seasonality is now considered to have been a major source of selective pressure during the mid/late Miocene, when the human ancestral line split from that of the great apes. During the late Miocene, evergreen tropical forest was gradually replaced by a diverse mosaic of woodland, bushland and savannah grassland, in which both rainfall and the availability of resources became more seasonal (Foley 1993). Wells (2006) suggests that the human capacity for ready fat deposition began to emerge during this period, and that moderate food shortages associated with subsequent seasonality in Africa, where anatomically modern humans evolved, would have maintained the pressure for humans to be able to store fat.

A unique pressure on energy resources in humans is our relatively very large brain size. The brain requires a great deal of energy and a constant supply of it, in the form of glucose. As the brain became relatively larger in the hominin evolutionary line, these needs were met in a variety of ways, one of which was probably by increasing fat stores (Kuzawa 1998). The energetic costs of growing a large brain are also very high. Human infants need to lay down considerable fat to sustain them before breastfeeding is properly established, and during the nutritional and infectious disease challenges often associated with weaning (Kuzawa 1998).

There are marked sex differences in body composition amongst foragers, with men having between 5 and 15 per cent body fat, and women between 20 and 25 per cent (Eaton and Eaton 1999b). The patterning of fat is also different, with women having more subcutaneous fat on the hips and limbs compared to the trunk (Brown and Bentley-Condit 1998). Greater female fatness, as well as female fat patterning, is probably related to the energetic demands of pregnancy and lactation (Rebuffé-Scrive *et al.* 1985). There is evidence that oestrogen facilitates efficient fat storage in premenopausal women (O'Sullivan *et al.* 2001) and that fat deposition is favoured in the femoral (thigh) fat depot, while during breastfeeding lipid mobilisation increases in this depot (Rebuffé-Scrive *et al.* 1985).

Despite having proportionately more fat than other species, measures of weight in relation to height for hunter–gatherers show that, compared to westerners, they are typically lean (Jenike 2001). For example, mean body mass index (BMI, calculated as weight in kg/height in metres^2) in the Hadza of Tanzania is 20, at the low end of the normal range, defined as between 18.5 and 24.9 (Sherry and Marlowe 2007). Skinfold measurements, which allow more direct assessment of the amount of adipose tissue on the body, suggest that hunter–gatherers have half the body fat typically found in industrialised westerners (Eaton and Eaton 1999a). Body fat levels remain stable or decrease with age, in contrast to the increased obesity seen with age in industrialised societies (Eaton and Eaton 1999b). For example, medics visiting the Dobe !Kung in the late 1960s observed no obesity and 'no middle-aged spread' (Truswell and Hansen 1976).

As noted in Chapter 2, degenerative diseases associated with over-consumption and lack of physical activity are absent in modern-day hunter–gatherers, and are considered to have been absent before the advent of agriculture (Cohen 1989). Insulin sensitivity (the importance of which is outlined below) is high amongst habitual hunter–gatherers (Truswell and Hansen 1976; Cordain *et al.* 1997). Evidence from clinical examinations shows that they hardly ever develop atherosclerosis, the main pathology underlying coronary heart disease (Eaton and Eaton 1999b). In the Dobe ! Kung assessed in the late 1960s total serum cholesterol level was very low and blood pressure was low and did not rise with age (Truswell and Hansen 1976). This pattern of low blood pressure and no increase in blood pressure with age appears to be consistent across populations living at subsistence level, in contrast to the age-related increase in blood pressure in western populations, which was once considered normal (Pollard *et al.* 1991).

The 'diabesity epidemic'

There is a stark contrast between the situation outlined above and that in western populations today. Put simply, the very different environment we now inhabit leads to obesity because the human body evolved for the circumstances of hunter–gatherer life, not for those of affluent western society. Currently, our way of life is moving even further away from that experienced during the vast majority of our evolution and levels of overweight and obesity, which are already high, are rising rapidly. Because the disease most strongly associated with obesity is type 2 diabetes, the term diabesity epidemic is sometimes used to describe this phenomenon (Astrup and Finer 2000). In this section I describe the increase in rates of obesity and type 2 diabetes and consider the proximate causes for this increase.

The rise in obesity

The term obesity refers to excess body weight with an abnormally high proportion of body fat. Overweight is usually defined as a BMI of 25 to 29.9, and a BMI of 30 or above indicates obesity. An average BMI of over 25 is not usually observed amongst hunter–gatherers (Jenike 2001) and, in fact, such a population statistic did not emerge in any human population until very recently. Overweight is also unusual amongst subsistence agriculturalists, who have to work very hard to obtain their food and who experience, on average, greater seasonal weight losses than hunter–gatherers (Brown and Konner 1987). The only historical group showing evidence of obesity is the small rich elite of urban societies. For example, in the mediaeval period in Britain the elite of 'gentry, lords of the manor, merchants, knights and some residents of monastic institutions' had access to an increasingly diverse range of foods, some of them in large quantities (Roberts and Cox 2003, p. 245). We have evidence of this in the skeletal condition known as diffuse idiopathic skeletal hyperostosis, found in only a small percentage of the population during this period (Roberts and Cox 2003, pp. 245–246). Obesity probably remained confined to the elite group until well after the Industrial Revolution, which produced an economy that remained reliant on the hard manual labour of men, women and children. Data from veterans who fought for the Union Army during the US Civil War (1861–1865) who were measured between 1890 and 1900, when they were between 40 and 69 years old, showed a prevalence of obesity of only 3%–4% (Helmchen and Henderson 2004). Mean BMI was 23.2 in men in their forties and 22.9 in men in their sixties.

Large national surveys, which included measurements of height and weight, and thus allowed the calculation of BMI, began in the 1950s and 1960s. By this time the prevalence of obesity in the US was 13% amongst

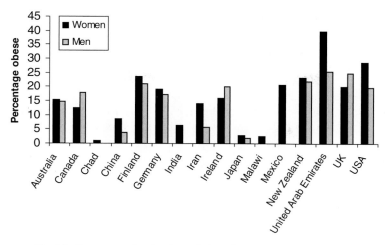

Fig. 3.2. Percentage of national populations that are obese (i.e. BMI > 29.9). Most recent data from the World Health Organization's Global Database on BMI. Note that surveys vary in terms of sampling and age ranges. Website: http://www.who.int/bmi/.

men and women aged 20–74 (Flegal *et al.* 1998). However, the really dramatic increase in excess weight in much of the western world happened at the end of the twentieth century. In the worst affected country, the United States, the prevalence of obesity increased dramatically during the 1990s (Mokdad *et al.* 1999) and more than half of all adults were classified as overweight or obese by the end of the twentieth century (Sowers 2003). The prevalence of obesity in adults in a number of industrialised countries is shown in Fig. 3.2.

Perhaps even more strikingly, childhood overweight and obesity have become common, and continue to increase rapidly in the United States (Strauss and Pollack 2001), the UK (Stamatakis *et al.* 2005) (Fig. 3.3) and other countries. In 1998 the prevalence of overweight or obesity in children in the United States was 22% among African-Americans and Hispanics and 12% among non-Hispanic whites, according to standard definitions used for children (Strauss and Pollack 2001). Obese children are at greater risk of becoming obese adults than are normal weight children (Robinson 2001) and there are serious physical and psychosocial health consequences associated with childhood overweight and obesity (Must and Colclough-Douglas 2001).

In 2005 a World Health Organization report suggested that there were over 300 million obese people and over 750 million overweight people in the world (WHO Global Infobase Team 2005). They estimated that around

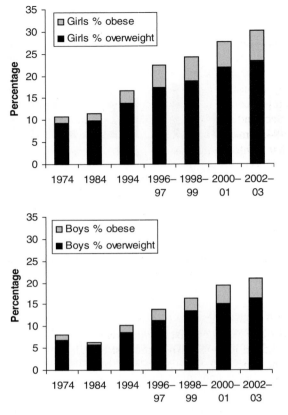

Fig. 3.3. Percentage overweight and obese 5–10-year-old children in England between 1974 and 2003, as defined by BMI according to standard cut-offs (Cole *et al.* 2000). Data from Stamatakis *et al.* (2005).

2.5 million deaths per year could be attributed to obesity in 2000. They also assessed the disease burden attributable to excess BMI around the world using Disability-Adjusted Life Years (DALYs), which express years of life lost to premature death and years lived with a disability, adjusted for the severity of the disability. Essentially, one DALY is equivalent to 1 lost year of healthy life. The WHO estimate was that the disease burden attributable to excess BMI worldwide was around 30 million DALYs in 2000.

These figures rely on measurements of BMI, but BMI does not differentiate between fat-free mass and fat mass. For example, body builders have a very low percentage of body fat, but their BMI may be in the overweight range because of their large muscle mass. Since it is the amount of fat that is the important issue here, it is also useful to examine more direct measures of fat where they are available. It is possible to assess fat directly using expensive techniques requiring specialised and clinic-based equipment, such as dual-energy X-ray absorptiometry, but this is not practical in large-scale surveys (Snijder *et al.* 2005). The most common measurement taken in surveys is subcutaneous skinfold thickness, usually at the triceps and/or subscapular skinfold as reported for hunter–gatherers on p. 34. Skinfold measurements in children in Scotland and England confirm that children became fatter between 1972 and 1994. Similarly, a review of data from several national surveys in the US found that skinfold thicknesses in US children have increased since 1960 (Eisenmann 2003).

Assessment of the patterning of fat on the body is also important. Excess abdominal fat, measured using waist circumference or the ratio of waist to hip circumferences or of waist to height, is particularly predictive of metabolic disturbances and cardiovascular and type 2 diabetes risk (Despres 2006; Wild and Byrne 2006), a fact that has recently been emphasised in health pro-motion strategies. Abdominal fat, especially visceral fat, acts differently from other fat tissue and is an important source of free fatty acids and metabol-ically active cytokines that affect the way in which the liver deals with glucose and fat (Despres 2006). Waist and hip measurements have been taken less frequently than height and weight in population surveys. However, data from the US show that the waist circumference of adults increased by 3.3 cm in men and 3.5 cm in women over the last decade of the twentieth century. Over the same time period the prevalence of abdominal obesity in US chil-dren, defined according to waist circumference, increased by nearly 70% (Li *et al.* 2006). Similarly, data from New Zealand show that, in addition to an increase in BMI between 1977 and 1997, there was a substantial increase in abdominal obesity (Wilson *et al.* 2001).

Type 2 diabetes

Increases in obesity over the past 30 years have been paralleled by a dramatic rise in the prevalence of type 2 diabetes, and the two are closely related (Seidell 2000; World Health Organization 2003). Cross-sectional data from the US show that overweight and obesity are strong risk factors for type 2 diabetes in men and women, and that this association is stronger than those of overweight and obesity with other diseases (Must *et al.* 1999). Most people with type 2 diabetes are overweight or obese (Wild and Byrne 2006). Earlier onset of type 2 diabetes is also associated with a higher BMI and increasing prevalence of overweight and obesity is the most important factor in the increasing number of younger people diagnosed with type 2 diabetes (Wild and Byrne 2006).

Obesity is strongly causally related to insulin resistance, which is the main underlying pathology of type 2 diabetes, as well as being an important risk factor for cardiovascular disease (Reaven 1988). The main role of insulin is to regulate levels of glucose in the body. It is secreted by the pancreas when blood glucose levels rise as a result of the consumption of carbohydrates. It prompts the uptake, storage and use of glucose by almost all tissues of the body, suppresses the production of glucose by the liver and is an important regulator of lipid metabolism. In people with insulin resistance the body, particularly the skeletal muscle and the liver, does not respond in the normal way to insulin. As a result, the pancreas secretes more insulin, leading to high insulin levels (hyperinsulinaemia) and this may regulate glucose levels successfully for a time. However, in the long run, when the pancreas is no longer able to produce sufficient insulin to overcome insulin resistance and maintain normal glucose levels, blood glucose levels become raised and, eventually, type 2 diabetes develops.

Although often initially unnoticed by the patient because it does not cause obvious symptoms in the early stages, type 2 diabetes is a serious disease. People with type 2 diabetes can suffer from a number of complications, including problems with blood supply to peripheral parts of the body and damage to sensory nerves (Pirart 1978). Foot ulcers, sometimes requiring foot amputation, and visual impairment are some of the most common consequences of this damage. People with type 2 diabetes also have a greater incidence of cardiovascular disease and kidney disease than the general population (Adler *et al.* 2000). Although type 2 diabetes is not commonly given as a cause of death, longitudinal studies following the mortality of people diagnosed with the disease show that they have a high death rate (Gu *et al.* 1998; Roper *et al.* 2001). The most common cause of death in those with type 2 diabetes in the UK and USA is heart disease (Gu *et al.* 1998; Roper *et al.* 2001). Poorly controlled diabetes, which is more common in poorer countries, brings higher risks of illness and death.

In the USA, the prevalence of type 2 diabetes increased over the 1980s and 1990s. By the early 1990s, 19% of adults in the USA aged 60 or over had diabetes (Harris *et al.* 1998). The number of new cases diagnosed each year also rose, doubling between the 1970s and 1990s in one community-based sample (Fox *et al.* 2006). In the west, even some children and adolescents now have type 2 diabetes (Kohen-Avramoglu *et al.* 2003). However, the largest numbers of estimated cases of diabetes in 2000 were in India (31.7 million) and China (20.8 million), followed by the USA (17.7 million) (Wild *et al.* 2004). The higher numbers in India and China are a result of the very large populations in those countries, but also reflect a rapid rise in diabetes prevalence in both countries.

Table 3.1. *Components of the metabolic syndrome, based on the definition of the International Diabetes Federation. HDL = high-density lipoprotein, LDL = low-density lipoprotein*

Component of metabolic syndrome	Explanation
Central obesity	Obesity where fat deposition is primarily around the trunk rather than the limbs
Insulin resistance	The body does not respond to insulin
Hyperinsulinaemia	Levels of insulin are high
Dyslipidaemia	High levels of triglycerides, low levels of HDL cholesterol, small dense LDL cholesterol particles
Hypertension	High blood pressure

For details of cut-offs for each component, see Alberti *et al.* (2005).

Type 2 diabetes and cardiovascular disease can be seen as the tip of an iceberg, which consists of a much larger number of people with metabolic syndrome (Zimmet 2000). This term is used to describe a constellation of related metabolic disturbances, the most important elements being abdominal obesity, dyslipidaemia (a harmful serum lipid profile), insulin resistance, and hypertension (Table 3.1). A number of different definitions of metabolic syndrome are currently in use, so that it is hard to define its prevalence, but a recent study, using one of the well-established definitions, found that in the USA prevalence ranged from 7% in people aged 20–29 years to 40% in people aged 60 or over (Ford *et al.* 2002).

Explaining recent increases in obesity and type 2 diabetes

Why have we witnessed such an explosion in rates of obesity and its consequences, especially type 2 diabetes, in the last 30 years? Unfortunately, the answer to this question remains incomplete, although contributory causes have been identified and are reviewed below. We do not have a complete answer partly because of methodological issues; it is difficult to measure energy intake and expenditure accurately. Since a small excess of energy intake sustained over time can lead to quite large gains in fat, this error is likely to confuse the interpretation of data. To complicate matters, it is possible that factors other than energy intake and expenditure play a role in causing obesity. Thus sleep duration has recently been identified as a possible determinant of obesity, perhaps via effects on energy intake and expenditure, but perhaps via other mechanisms. It is also possible that chronic inflammation may play a role in the development of obesity.

Diet

Perhaps surprisingly, energy intake has declined over the twentieth century in at least some western countries. Even over the last 30–50 years energy intake has declined in England and Canada (Prentice and Jebb 1995; Gray-Donald *et al.* 2000), although in the United States it appears to have been fairly stable over this time period (Arnett *et al.* 2000; Troiano *et al.* 2000). However, it is clear that there must have been an increase in energy imbalance, such that people have increasingly consumed more energy than they require in their daily activities.

One reason for the failure to maintain a balance between dietary intake and energy expenditure is thought to be a change in the macronutrient content of the diet, with a higher proportion of energy-dense foods and drinks. Consumption of foods high in fat and energy make over-consumption of energy more likely, partly because a large number of calories are consumed in a relatively small portion of food. In addition, because human appetite control systems evolved in very different environments, they do not recognise the large amounts of energy in high-fat and energy-dense foods and drinks, and thus do not down-regulate the quantity of food eaten in a way that would maintain energy balance (Mann 2004).

The consumption of this type of food is strongly associated with the consumption of more food outside the home, particularly in the USA, where the proportion of meals eaten outside the home, particularly in fast-food restaurants, increased greatly between the 1970s and 1990s (Briefel and Johnson 2004). This change has been driven by rising incomes and increased demand for convenience, partly associated with the increase in the number of women working outside the home. There has been an unprecedented growth in the number of commercial food outlets in the US since 1970, with especially dramatic growth in the fast-food industry (Kant and Graubard 2004). Furthermore, fast-food outlets, and other food manufacturers, increased their portion sizes in the 1990s in the phenomenon known as supersizing. For example, in a recent polemic, Critser (2003, p. 28) suggests that a portion of french fries from McDonalds provided 200 calories in 1960, but 610 calories in 2002. Between 1977 and 1996, portion sizes in the USA also increased substantially for salty snacks, desserts, soft drinks, hamburgers and Mexican food, both outside and inside the home (Nielsen and Popkin 2003).

Consumption of foods with a high glycaemic index, which cause a big rise in blood glucose levels, is higher now than ever before. As we have seen, raised glucose levels stimulate insulin secretion, promoting uptake of glucose by muscle and adipose tissue. As a result, insulin levels are higher after the consumption of foods with a high glycaemic index, and other hormone levels are also affected. The initial period of high levels of blood glucose and insulin is usually followed by reactive hypoglycaemia and hormone changes designed to normalise glucose levels again. This pattern of metabolic changes may promote excessive food intake, dysfunction of pancreatic beta cells and dyslipidaemia. There is general agreement that habitual consumption of high glycaemic index foods is likely to increase risk of obesity, insulin resistance, type 2 diabetes and coronary heart disease (Ludwig 2002), although some are not convinced (Pi-Sunyer 2002). In general, most refined starchy foods eaten in western societies, including bread and starchy root vegetables such as potatoes, have a high glycaemic index, whereas non-starchy vegetables, fruit and legumes tend to have a low glycaemic index (Ludwig 2002). Consumption of flour increased considerably during the last 20 years of the twentieth century in the USA and total dietary sugar intake increased from 55.5 kg per person per year in 1970 to 69.1 kg in 2000 (Cordain *et al.* 2003; Lieberman 2003). Refined grain products with a high glycaemic load now supply 20% of the energy in the typical US diet and high glycaemic load sugars supply 16% of the energy (Cordain *et al.* 2003).

Conversely, consumption of dietary fibre declined over the twentieth century in western countries with the increased use of refined grain. This decline was particularly marked in the United States, where fibre consumption is well below recommended values (Gross *et al.* 2004; Cordain *et al.* 2005). A high intake of dietary fibre is considered to protect against weight gain and the development of obesity (World Health Organization 2003), perhaps because fibre promotes satiety and slows down the processes of starch digestion or absorption, leading to relatively small glucose and insulin responses (Liu *et al.* 2003). In one large US study, women who consumed more whole grains, with a high fibre content, consistently weighed less than did women who consumed fewer whole grains, and follow-up studies showed that, over a period of 12 years, a decrease in dietary fibre intake was associated with increased weight (Liu *et al.* 2003). Low dietary fibre content is probably also a risk factor for type 2 diabetes. Thus the decreased per capita consumption of dietary fibre was strongly associated with the increased prevalence of type 2 diabetes in the United States over the twentieth century, even in analyses controlling for energy intake (Gross *et al.* 2004).

Snacking behaviour has also changed in the USA, with more snacks eaten, more energy coming from snacks as opposed to meals, and an increased energy density of snacks (Zizza *et al.* 2001). Snackers appear to have a higher total energy intake than those who do not snack (Zizza *et al.* 2001). It has also been hypothesised that snacking (sometimes called grazing) is, in itself, harmful because more time is spent in a post-prandial metabolic state and this may favour fat deposition (Frost and Dornhorst 2001). However, it is not snacking *per se* that is harmful. We have seen (above) that Australian Aborigines, for example, 'would eat snacks throughout the day while hunting and gathering' (O'Dea 1991). Such snacking makes sense because it means that not all food has to be carried, and it did not result in obesity. It is the type of food consumed as snacks that is important. In contrast to the snacks available to the Australian Aborigine – 'grubs, fruits, gum, honey ants and sugar-bag (honey from the wild bees)' (O'Dea 1991), those most commonly eaten by young American adults are desserts (ice cream, apple pie, cookies and cakes), sweetened drinks (mostly sweetened soft drinks), alcoholic beverages, and salty foods (mainly high fat salty snacks such as potato chips (known as crisps in the UK)) (Zizza *et al.* 2001). Sweetened drinks probably have a particularly important role in the rise of obesity – research has shown that children and women with a high consumption of soft drinks with high sugar content are likely to be overweight (Ludwig *et al.* 2001; Schulze *et al.* 2004).

Most of the evidence for changes in dietary behaviour comes from the USA, which has the biggest obesity problem. However, many of these changes are spreading to other parts of the world, as evidenced by the presence of US fast-food chains around the world. For example, Lieberman (2003) cites the 2000 annual report for Coca-Cola, whose non-alcoholic drink market sales are global; 30% in North America, 26% in Latin America and 21% in Europe and Eurasia. The globalisation of world food markets and of consumption patterns is likely to ensure that diets worldwide become increasingly similar to that in the USA. The impact of such changes in less developed countries is discussed in Chapter 9.

There have been signs of positive changes in diet in some populations in recent years. For example, positive changes in serum lipid profiles and a decrease in blood pressure in Finland between 1972 and 1997 have been attributed to a documented decline in the proportion of energy derived from dietary fat over this period (Vartiainen *et al.* 2000). The main causes of this change are thought to have been less use of butter and a shift from using whole milk to using low-fat and skimmed milk. A decline in the percentage energy from fat in the diet was also observed in British adolescents between 1990 and 2000 (Fletcher *et al.* 2004). It seems that messages about the importance of fats, and particularly saturated fats, are being heeded in some western populations.

Energy expenditure

Physical activity is an essential part of the equation determining energy balance. In a large Finnish study, low levels of physical activity were iden- tified as a more important risk factor for excess weight gain than any features of the habitual diet (Rissanen *et al.* 1991). Similarly, correlational analyses of diet and proxy measures of physical activity with obesity over time and across social classes using British data suggest that inactivity is a more important determinant of obesity than diet (Prentice and Jebb 1995). It has been suggested that waist circumference is more sensitive to changes in physical activity than BMI. For example, if an individual reduces his or her level of physical activity, muscle may be gradually replaced by fat, which will have less effect on overall body weight than on waist circumference, since the abdomen will be relatively little affected by loss of muscle and more affected by fat gain (Snijder *et al.* 2005). Physical activity is also important in its own right in protecting against diseases of insulin resistance. For example, low cardiorespiratory fitness is a strong predictor of metabolic syndrome risk in women and men, independent of BMI (LaMonte *et al.* 2005).

In the latter half of the twentieth century activity patterns probably changed more dramatically than eating behaviour. Work became more sedentary with the advent of automation and most members of affluent societies do not now engage in any hard occupational activity; physical activity is now usually a leisure-time activity. At the same time, sedentary recreation, such as watching television and using computers, has become much more common (McMichael 2001), as have energy-saving devices such as cars, lifts and television remote controls. In England the average person watched 13 hours of television a week in the 1960s, but over 26 hours a week in the 1990s (Prentice and Jebb 1995). In the USA today cars are needed in everyday life because most people do not live within walking distance of work or shops, and time spent commuting adds to the difficulty of finding time for recreational physical activity (Brown and Krick 2001). As a consequence, the prevalence of a sedentary lifestyle in the USA gradually increased towards the end of the twentieth century (Dietz 1996).

It is not surprising, then, that many people in affluent nations are now extremely sedentary. Less than 10% of American women and men engage in regular and vigorous physical activity (Catanese *et al.* 2001) and levels of physical activity in Europe are also low, with around a third of the population reporting that they engaged in no 'physical exercise' during a typical 2-week period (reported in Catanese *et al.* 2001). Lack of time is an important barrier to physical activity undertaken during leisure time, as indicated by a British study showing that working long hours and having children was associated with less leisure time exercise in the working age population (Popham and Mitchell 2006).

There is particular concern about inactivity in children. As we have seen, hunter–gatherer children are very active, but children in affluent societies today are remarkably inactive. In the USA children spend more time watching television, on average, than in any other activity except sleep, including time spent in school (Brown and Krick 2001), and there is strong evidence that the more time children and adolescents spend watching television, the fatter they become (Kaur *et al.* 2003). Children from many other parts of the world also watch television for several hours per day on average (e.g. in the United Arab Emirates, Henry *et al.* 2004). Sedentary patterns are established at a very early age. A Scottish study which measured energy expenditure in children at the ages of 3 and 5 years indicated that they were already living a sedentary lifestyle, with children typically spending only 20–25 minutes per day in moderate to vigorous physical activity (Reilly *et al.* 2004).

Sleep

An association between short sleep duration and elevated risk of obesity has been established recently (Reilly *et al.* 2005; Vorona *et al.* 2005; Taheri 2006). Short sleep duration has also been posited as a risk factor for insulin resistance and type 2 diabetes (Spiegel *et al.* 2005). It may be that sleep loss directly affects metabolism. For example, short sleep duration is associated with reduced levels of leptin and increased levels of ghrelin in morning blood samples (Taheri *et al.* 2004). These hormones are key regulators of appetite and the changes observed in association with short sleep duration are likely to increase appetite. Sleep loss may also lead to increased dietary consumption because of increased waking hours and thus opportunity to eat, and to a decrease in physical activity levels because of tiredness (Taheri 2006). However, it is not yet clear that sleep loss causes decreased physical activity levels. It is perhaps more plausible that low physical activity levels cause short sleep duration.

There is evidence that sleep duration in western societies is now particularly low. Data from Switzerland show that, for children, total sleep duration fell between 1974 and 1993 because of increasingly later bedtimes but unchanged waking times (Iglowstein *et al.* 2003). Similarly, a survey undertaken in the USA found that the proportion of young Americans reporting less than 7 hours sleep per night rose from 16% in 1960 to 37% in 2002 (National Sleep Foundation 2002). Research has also suggested that young adults in the USA typically sleep less than they need to, resulting in daytime fatigue (Bonnet and Arand 1995).

There has been very little research on sleep in non-western societies and it is not currently possible to estimate how long humans might have slept in any given 24-hour period for most of our evolutionary history. However, Worthman and Melby (2002) note that physical workload appears to be correlated with sleep duration in some non-western societies. They cite the example of hard-working horticulturalist Gebusi women from highland Papua New Guinea, who typically sleep for 10 hours at night. They also note that, for hunter–gatherers, who have little need to schedule their work, times of falling asleep vary widely within and among people and there is great fluidity in sleep–wake patterns. As a consequence, in hunter–gatherers and many non-western societies sleep tends not to be consolidated into one night-time bout as it is in western societies. Thus patterns of sleep are strikingly different from those seen in western societies.

It seems clear that the western context must moderate the effects of sleep duration. Given the absence of excess adiposity for most previous generations of humans, the correlation between sleep duration and obesity is probably novel. It would, however, be helpful to understand more about why hunter–gatherers sleep in the way that they do in order to understand more about how the human body is adapted to particular sleep patterns.

The role of inflammation

Inflammation occurs in response to infection and injury and involves the production of cytokines and oxidant molecules, which help create a hostile environment for pathogens and repair damaged tissue. We now know that adipose tissue, in particular abdominal adipose tissue, secretes a range of inflammatory factors. It is unclear, however, whether adiposity is an important cause of inflammation or whether chronic inflammation may pre-cede systemic insulin resistance and adiposity. In any case, there is probably a positive feedback cycle between inflammation and adiposity (Haffner 2003; Wells 2006). As connections between infection, inflammation and apparently non-communicable diseases continue to be explored, it has been suggested that infections play a dominant role in the causation of many, if not most, chronic diseases (Ewald 2002). Others have suggested that a reduction in lifetime exposure to infectious diseases, and the inflammation they induce, has made an important contribution to the historical decline in mortality in old age (Finch and Crimmins 2004). This field is underdeveloped as yet, but is likely to lead to valuable new insights into the aetiology of western diseases.

Cardiovascular disease

The most important forms of cardiovascular disease today are coronary heart disease (also known as ischaemic heart disease), which is caused by a block in the supply of blood to the muscle of the heart, and stroke, which may be caused by the same mechanism or by rupturing of a blood vessel in the brain. Coronary heart disease was the largest single cause of disease burden worldwide in 1990 (Murray and Lopez 1997). As we have seen (p. 40), it is the main cause of death in those diagnosed with type 2 diabetes in the UK and USA.

The main underlying pathology of coronary heart disease and the first form of stroke is atherosclerosis, the formation of plaques with a cholesterol-rich lipid core within arteries supplying blood to the heart and the brain. The earliest manifestations of atherosclerosis are fatty streaks in the artery walls, seen in most children in western societies, which may progress to become plaques from the third decade of life onwards (Fuster *et al.* 1992a). These atherosclerotic plaques narrow the lumen of the artery and can trigger a blood clot, resulting in complete obstruction of the artery (Fuster *et al.* 1992a; Fuster *et al.* 1992b). When coronary arteries are affected in this way, the loss of blood supply, and thus oxygen supply, to the heart results in coronary heart disease. Ischaemic strokes are often precipitated by the same mechanism in the brain. Strokes due to bleeding in the brain are usually associated with hypertension (high blood pressure). Hypertension is often found in association with atherosclerosis and probably contributes to the atherosclerotic process by causing damage to the arterial wall, which precipitates local atherosclerosis (Chobanian and Alexander 1996). Hypertension appears primarily to be caused by increased resistance in peripheral blood vessels as a result of widespread constriction of the arterioles and small arteries, and an increase in blood volume can also be a contributory factor (Julian and Cowan 1992).

Risk factors for cardiovascular disease

Harmful serum lipid profiles, which are strongly associated with obesity and metabolic syndrome, contribute to the process of atherosclerosis. High levels of low density lipoprotein (LDL) cholesterol are more important than total cholesterol levels in increasing risk, whereas high levels of high density lipoprotein (HDL) are protective against atherosclerosis. Diet, physical activity and obesity clearly contribute importantly to lipid levels. The most deleterious component of the diet is saturated fat, mainly derived from animals. Further, lipoproteins are more likely to be deposited in the arteries if they are oxidated. Dietary antioxidants such as vitamins C, E and beta-carotene, which are found in fruit and vegetables, probably have a protective role because they help prevent oxidation of LDL (Van Poppel *et al.* 1994; Stampfer and Rimm 1995). The only fats that appear to have clear beneficial effects are omega-3 fatty acids, as noted above. Physical activity is important not only in determining long-term energy balance, and therefore weight gain, but because it has independent beneficial effects on lipid profiles, coagulation and thrombosis, blood pressure and insulin sensitivity (Kannel 1987; Bassuk and Manson 2005).

Obesity is also an important risk factor for hypertension (Must *et al.* 1999; Wild and Byrne 2006), as is the consumption of sodium, usually as salt. Sodium is thought to raise blood pressure by causing an increase in blood volume. Consumption of sodium has been shown to correlate with blood pressure both across and within populations (Elliott *et al.* 1996). Diets in affluent countries contain considerably more sodium than most traditional diets, with much of it coming from salt added to processed food and bread. The ratio of sodium to potassium in the diet is also important for hypertension risk (Young *et al.* 1995).

Smoking is a very important risk factor for cardiovascular disease. It increases the risk of blood clotting, causes high blood pressure and leads to an adverse lipid profile (Kannel 1987; Muscat *et al.* 1991; De Boever *et al.* 1995). Psychosocial stress and depression are also risk factors for cardiovascular disease in western societies, as we shall see in Chapter 8. Chronic inflammation has been linked with coronary heart disease and it is likely that inflammation is an important common link between obesity, type 2 diabetes and coronary heart disease (Ridker 2002). Specific infections associated with the development of coronary heart disease include most notably the bacterial infections *Helicobacter pylori* and *Chlamydia pneumoniae*, although the evidence of a causal link is equivocal (Ridker 2002; Mussa *et al.* 2006). Thus the risk factors for type 2 diabetes and for cardiovascular disease, especially coronary heart disease, are similar, although not identical (Mann 2002).

Trends in mortality and morbidity from cardiovascular disease

Trends in mortality and morbidity from cardiovascular disease have been rather different from those for obesity and type 2 diabetes, partly because there are a number of cardiovascular diseases, with differing risk factors. In rough summary, in the west there appears to have been a reduction in some forms of cardiovascular disease between the mid-nineteenth and mid-twentieth centuries, an increase in coronary heart disease and stroke over the first half of the twentieth century, and then a decline in coronary heart disease and stroke over the second half of the twentieth century.

Different analyses of the same data for 165 populations from around the world (but with a bias towards Europe and more affluent nations) found that age-specific cardiovascular mortality declined in both sexes between 1861 and 1964 (Preston 1976; Gage 2005). Light is thrown on these findings by information from veterans of the Union Army, who fought in the American Civil War between 1861 and 1865. This information is available from the records of the Union Army pension programme, the most widespread form of assistance to the elderly before Social Security began in the USA. In the early years of the twentieth century it benefited an estimated 25% of the US population older than 64 (Costa 2000). Men underwent detailed medical examinations for the pension programme, and these records have been linked to census data (providing information on occupation, for example) and military service records (providing information on infectious diseases and injuries suffered, amongst other things). These data can be compared with data collected late in the twentieth century from army veterans, although, importantly, it is only possible to compare conditions that could be diagnosed accurately for the Union Army veterans. Using these data, Fogel and Costa (1997) found that men older than 64 in 1910 were much more likely to suffer from heart, respiratory, musculoskeletal and digestive disorders than their counterparts at the end of the twentieth century. This difference was not small; Costa (Costa 2000) reports that the prevalence of irregular pulse rates, murmurs, and valvular heart disease declined by 90% from 1900–1910 to the 1970s-1990s.

How can we explain these trends? Why might the risk of developing some forms of cardiovascular disease at any given age decline as people start to suffer less from undernutrition and infectious disease? Costa (2000) suggests, after careful analysis, that nearly a third of the overall decline in chronic conditions can be explained by occupational shifts from manual to white-collar and more sedentary work and that nearly a fifth can be explained by the reduction in infectious disease. She points out that manual work in the nineteenth century was very hard, often involving exposure to dust, fumes and animal and industrial pollutants, and so it is not surprising that manual workers at this time suffered from poor health. She also showed that, for Union Army veterans, the experience of acute respiratory infections while in the army increased the probability of chronic diseases in later life. This finding fits with evidence discussed above linking infections and inflammation with cardiovascular and other diseases. In particular, rheumatic heart disease was a much more important form of cardiovascular disease in the nineteenth than in the twentieth century, and is caused by earlier experience of rheumatic fever.

Costa (2004) also links anthropometric changes to the decline in chronic disease amongst male army veterans in the USA. Here, she draws on anthropometric measurements of Union soldiers taken during the 1860s. In the 1860s military men were shorter and lighter than in 1950 or 1988. In analyses controlling for BMI, waist–hip ratio was significantly greater in the 1860 sample than in the 1950 or 1988 military. This indicates that there was greater central patterning of fat in the 1860s sample. This is an intriguing finding, given the recent work on links between lifetime infection, inflammation, abdominal obesity and type 2 diabetes/cardiovascular disease risk (see p. 43 and p. 45), and also between slow growth in early life and abdominal obesity (see p. 66).

However, not all forms of cardiovascular disease could have been diagnosed in the Union Army veterans, and it seems clear that the age-specific risk of some forms of cardiovascular disease increased in western countries over the first 50–60 years of the twentieth century. Davey Smith and Marmot (1991) identified a 'real increase' in cardiovascular mortality in England and Wales between 1920 and the 1960s because of a rise in coronary heart disease, despite a decline in other forms of cardiovascular disease, including rheumatic heart disease and valvular heart disease. The rise in coronary heart disease was bigger in men than in women (Davey Smith and Marmot 1991). Similarly, there was a rise in age-specific mortality from coronary heart disease between 1925 and the early 1950s in the USA (Slattery and Randall 1988). In England and Wales between the 1930s and 1990s, there was a rise in mortality from stroke caused by a blood clot cutting off the supply of oxygen to part of the brain, but a decline in mortality from stroke caused by bleeding into the brain (Lawlor *et al.* 2002). Thus it appears that forms of cardiovascular disease associated with atherosclerosis and high levels of dietary fat intake and harmful serum lipid profiles increased during much of the twentieth century.

The changes in rates of coronary heart disease did not take place uniformly across different socio-economic groups. For example, mortality rates from coronary heart disease were higher for men of higher socio-economic status and showed no socioeconomic gradient in women in Britain between 1921 and 1951, but by 1971 there were higher rates in both men and women of lower socio-economic status (Marmot *et al.* 1984; Davey Smith and Marmot 1991). Thus coronary heart disease was initially largely associated with relative affluence, but as poorer sectors of society began to gain access to high fat diets and related risk factors, heart disease followed.

Subsequently in the west there was a downturn in mortality from cardiovascular disease, including both coronary heart disease and stroke, in the second half of the twentieth century (Caldwell 2001). Wilmoth (2000) notes that, in the United States 'from 1950 to 1996, age-adjusted death rates for these two causes declined by more than half (by 56% for heart disease, and by 70% for stroke)'. In Britain the decline in coronary heart disease in men began later, in the 1970s (Davey Smith and Marmot 1991). Contributory factors include, most importantly, a decline in cigarette smoking amongst adults, as well as a reduction in the consumption of saturated fat, increasing medical control of hypertension and better treatment of heart disease and stroke (Slattery and Randall 1988; Wilmoth 2000). Average blood pressure levels in western and other European populations have also been in decline (Tunstall-Pedoe *et al.* 2006), although it is unclear why. Possible reasons include changes in consumption of saturated fat, alcohol and sodium. It remains to be seen how the increasing rates of obesity and insulin resistance will affect mortality from coronary heart disease in the west in the future.

In poorer countries there is a continuing rise in cardiovascular mortality, which Yusuf *et al.* (2001a) attribute largely to the impact of urbanisation, with concomitant changes in diet, energy expenditure and a loss of traditional support mechanisms, as well as to more general dietary changes that have not been confined to urban populations. Increases in the use of tobacco are also important (Murray and Lopez 1997). We can think of this process as one of westernisation, while changes in diet have been referred to as the nutrition transition (Popkin 2001). For example, in China between 1982 and 1992 the proportion of energy provided by fat and the contribution of animal products to the diet grew and the consumption of carbohydrates declined. Similar changes have occurred in Korea, India, Indonesia, the Philippines and several other countries, leading to increasing rates of obesity, higher blood pressure, cholesterol and glucose levels and a decrease in insulin sensitivity (Yusuf *et al.* 2001b) (see pp. 157–163).

Other related diseases

More and more diseases have been linked to obesity and insulin resistance, some of which are considered in greater detail in later chapters, including reproductive cancers (see Chapter 5), polycystic ovary syndrome (see Chapter 6) and asthma (see Chapter 7). High levels of obesity and insulin resistance in western populations are also thought to contribute to higher risks of many other cancers, including those of the colon, kidney, oesophagus, pancreas, gallbladder and liver, and it has been estimated that 15–20% of all cancer deaths in the United States can be attributed to overweight and obesity (Calle and Kaaks 2004). The development of these cancers is affected by hyperinsulinaemia, an associated increase in the bioavailability of steroid hormones (see Chapter 5) and localised inflammation (Calle and Kaaks 2004).

There is also increasing evidence for strong associations between obesity, type 2 diabetes, cardiovascular disease, and their risk factors, and dementia. For example, a large longitudinal study in the USA showed that obesity in mid-life, as assessed by BMI and skinfold thicknesses, was a strong predictor of dementia of all kinds around 30 years later (Whitmer *et al.* 2005). Vascular dementia, one of the two most common types of dementia in affluent populations, is often caused by strokes (Grossman *et al.* 2006). More recently it has become clear that Alzheimer's disease, the other common type of dementia, is closely related to cardiovascular disease. Those with cardiovascular disease are more likely to experience cognitive decline or

Alzheimer's disease (Stampfer 2006). People with type 2 diabetes are also at greater risk, possibly because the vascular disease that often accompanies diabetes may affect blood vessels in the brain, or perhaps because of direct effects of insulin on the brain (Stampfer 2006).

Summary

Humans store a lot of fat compared to other animals that evolved on the tropical savannah. Selective pressures favouring fat deposition included the uniquely large relative brain size of humans and the need to survive seasonal environments. Humans evolved to eat wild plants and animals and foods with a low glycaemic load, not highly processed, energy-dense foods with a high glycaemic load. During most of human evolutionary history, people had to work fairly hard to make a living and were certainly much more physically active on a day-to-day basis than are the largely sedentary members of affluent societies today. The consequences of this mismatch between lifestyle and human biology are high rates of obesity, and associated diseases, including type 2 diabetes and some forms of cardiovascular disease. Dramatic changes in both diet and physical activity during the final years of the twentieth century and continuing into the twenty-first century have made the mismatch much greater than ever before.

4   *The thrifty genotype versus thrifty phenotype debate: efforts to explain between population variation in rates of type 2 diabetes and cardiovascular disease*

In 1974 a seminal paper drew attention to the very high rates of type 2 diabetes in many native populations of the Americas, Greenland, Polynesia, Micronesia and Melanesia (West 1974). West suggested that diabetes was probably uncommon in these groups prior to 1940, and noted that rates were still very low in those least touched by market economies. The high rates of diabetes in these populations were initially most often ascribed to a genetically based susceptibility to the development of obesity and associated diseases on the adoption of a western way of life. This explanation is commonly known as the 'thrifty genotype' hypothesis. An alternative explanation is that the susceptibility of these populations lies solely in the rapid changes of lifestyle they have experienced. Thus, it has been suggested that those born into a relatively poor environment may, if they later encounter a more western environment, be vulnerable to the development of obesity-related diseases. This latter explanation is sometimes referred to as the 'thrifty phenotype' hypothesis, to highlight its contrast with the thrifty genotype hypothesis. In this chapter I describe the populations best known for their very high rates of type 2 diabetes and cardiovascular disease, and then go on to consider in detail the competing explanations offered for these high rates of disease. I focus particularly on the debate about the causation of type 2 diabetes.

The problem: populations identified as having unusually high rates of type 2 diabetes/cardiovascular disease

Below, I describe several populations known for some time to have a markedly high prevalence of type 2 diabetes. Sometimes, crude prevalence

50

figures are given and sometimes prevalences are age standardised or age and sex standardised, so that the demographic structure of the population being studied is taken into account. Whilst the same method of standardisation is used within studies, this is not always the case across studies and cross-study comparisons must therefore be treated with caution. To provide some kind of benchmark, in 1995 the unadjusted prevalence of type 2 diabetes in adults aged over 20 years was estimated to be 6% in affluent countries and 4% worldwide (King *et al.* 1998).

Prior was among the first to note the high rates of obesity and type 2 diabetes in the islands of the South Pacific (Prior 1971). He led a research group which surveyed Polynesian populations on the 1960s and found very high rates of obesity and diabetes on some islands. In contrast, the more remote islands, less influenced by external economies, had very low rates of obesity and type 2 diabetes. His team also found that, in New Zealand, people of Māori origin had much higher rates of obesity and type 2 diabetes than people of European origin. Later, they investigated the health effects of migration to New Zealand by some of the indigenous Polynesian population of the island of Tokelau. In New Zealand the migrants became much fatter than did those left on Tokelau and the prevalence of type 2 diabetes approximately doubled, to an age-standardised prevalence of 11% in women aged over 20 (Stanhope and Prior 1980).

Massive adiposity and a high risk of type 2 diabetes was also observed in another group of Polynesians, urban Samoans, in the 1970s (Zimmet *et al.* 1981). This phenomenon was investigated further by a group of researchers who made use of the fact that some Samoans lived quite traditionally in what was then Western Samoa (now the nation of Samoa), others lived in what they termed modernising environments, for example, in American Samoa, and yet others lived in what they called modernised environments, such as California (Baker *et al.* 1986). Obesity and diabetes increased in a stepwise fashion from the traditional, to modernising, to modern environments. Subsequently, levels of obesity have continued to rise in women, men and children in Samoa (Keighley *et al.* 2007).

Similar findings have been made on other Pacific Islands. For example, a representative survey of Tonga in 1998–2000 found an age-standardised prevalence of diabetes of 15%, compared to a previously reported figure of 7.5% in 1971 (Colagiuri *et al.* 2002). The majority (80%) of Tongans with diabetes according to the standard tests performed as part of the survey did not know that they had the disease, leaving them at much increased risk for potentially fatal complications such as coronary heart disease, renal disease and diabetic foot.

The nation, as opposed to population, with the highest rate of diabetes in the world is Nauru, a small, very isolated Pacific island, only 12 miles in circumference, usually classified as Micronesian. For some time it had one of the highest per capita incomes in the world because of its phosphate industry; a high-grade phosphate ore is found over most of the island. This ore was extracted throughout the twentieth century, although it was only from 1966 that the Nauruans received significant payment for the right to extract the ore, from which time they became very affluent (McDaniel and Gowdy 2000). Zimmet and colleagues surveyed 12% of the island's population in 1975 and found that 34% of those over 15 were diabetic, while 11% were borderline diabetic (crude prevalence rates) (Zimmet *et al.* 1977). There was also a high prevalence of obesity. In 1987 the age-standardised prevalence of diabetes in those aged over 20 was around 34% (Dowse *et al.* 1991).

Probably the population most famous for its very high rates of obesity and type 2 diabetes is the Pima, a Native American group living in the south-west of the USA. Between 1965 and 1975, Pima adults aged over 35 and living in the USA had an age–sex adjusted prevalence of type 2 diabetes of 40–50% (Knowler 1978). Diabetes incidence (the proportion of the population developing diabetes each year) was the highest known, and was 19 times that in the predominantly white population of Minnesota. Very high levels of obesity, particularly abdominal obesity, and of type 2 diabetes have more recently been reported in other Native American populations (Young *et al.* 2000). For example, the prevalence of type 2 diabetes in one of the largest Algonquin communities of Quebec in Canada was about the same as that observed in Pima women (Delisle *et al.* 1995). The Oji-Cree, who live in a remote area of Ontario in Canada, also have astonishingly high rates of diabetes, reaching 80% among women aged 50–64 years (Harris *et al.* 1997, see also Box 4.1 and Figure 4.1). The few longitudinal studies in native north American populations have shown a rapid increase in the prevalence of obesity (predominantly abdominal) and diabetes over the past two decades (Young *et al.* 2000). Cardiovascular disease, particularly coronary heart disease, is now also more common amongst Native Americans in the USA than in the general population, despite initially being surprisingly low in the Pima (Howard *et al.* 1999). It is very clear, however, that it is only in those living a more western lifestyle that these problems emerge, since Pima living in Mexico and other Native American groups in South America do not show high rates of either type 2 diabetes or cardiovascular disease (Spielman *et al.* 1982; Schulz *et al.* 2006).

Box 4.1. The Oji-Cree of Sandy Lake, Ontario in Canada have been intensively studied because of their very high rates of diabetes. The account given here is taken from a paper by one of the researchers who led this work (Hegele) and a medical practitioner (Bartlett) who practised in the district which included the Sandy Lake area in the 1940s (Ray and Stevens 1995; Hegele and Bartlett 2003). Adapted with permission from the *Canadian Journal of Diabetes* 57: 257–261.

Many aspects of the traditional way of life in Sandy Lake were still intact in the 1940s. Western cultural influences were limited, partly by physical barriers – access to the region at that time was mainly by canoe and aeroplane. The people were nomads. In the summer, they lived in tents by the side of the lake and fished and in the winter, they dispersed to individual trap lines and lived in single-room log cabins. They had a modest caloric intake and relied on a narrow range of food sources. The main sources of protein were fish and rabbit. People remembered times when winter food supplies ran out altogether, leading to deaths from starvation. In the early twentieth century it was not uncommon for the people to walk up to 100 km/day. For longer distances, canoes were used in summer and dog sleds in winter, both requiring considerable physical effort. From the 1960s onwards access by road in the winter, across the frozen lake, became possible. Outboard motors were introduced to boats, followed by cars and snowmobiles. Television also brought about dramatic changes as indicated by the following experience of an elder: 'Preceding the days of television, children had free range all over the village in the evenings. In the dark, their voices and laughter could be heard as they ran and played. The shouts of the children were night music until they went inside, exhausted. On the night that television came, Ennis went outside in the evening darkness and was greeted with the strange emptiness of silence. Only the pathetic howls of a few lonely dogs rang in the quiet evening air. The glue of television has adults and children watching an alien world every night'. (Ray and Stevens 1995, cited by Hegele and Bartlett 2003).

Diabetes has also reached epidemic proportions among Australian Aboriginal people (Daniel *et al.* 2002). In one community in Central Australia in around 1990, 75% of women and 51% of men aged over 35 years were overweight or obese and around 30% (a crude prevalence rate) of survey participants in this age group had diabetes (O'Dea *et al.* 1993). The risk of diabetes in those who were aged between 25 and 64 was 11 times greater than in people from the same age group in the general

Fig. 4.1.   Summer life for the Oji-Cree in Sandy Lake, Ontario, Canada in the 1940s.
People lived in tent settlements close to the lake and fished for food. Reproduced with
permission from Hegele and Bartlett (2003).

Australian population. There is also greater abdominal obesity in Aborigines compared to people of European origin (Guest *et al.* 1993) and
Australian Aborigines have high rates of mortality from cardiovascular
disease (Thomson 1991).

Migrants to the affluent west from some of the poorest parts of the world
were also shown to have very high levels of type 2 diabetes from about the
late 1980s. The best-known example is probably populations of South Asian
origin (Misra and Vikram 2004). The trend for people of South Asian origin
to be more obese is less consistent, but it is clear that they tend to show more
abdominal obesity and a greater percentage of body fat than the general
population in countries like the UK and USA (Misra and Vikram 2004).
People of South Asian origin in the UK also have a very high risk of coronary
heart disease (McKeigue *et al.* 1991), but not of hypertension (Lane *et al.*
2002). In the poorest rural areas of South Asia these problems are far less
evident, although even in rural areas there are signs that the number of
diabetics is rising fast (Hussain *et al.* 2005; Chow *et al.* 2006).

In the UK there is also a raised prevalence of type 2 diabetes in people of
African or Afro-Caribbean origin, but, perhaps surprisingly, a markedly low
prevalence of coronary heart disease (Tillin *et al.* 2005; Forouhi and Sattar

2006). Favourable patterns of lipid levels are also seen in Afro-Caribbean men in the UK, in line with these relatively low rates of coronary heart disease (McKeigue 1996). Afro-Caribbean men seem to have less abdominal obesity than white men in the UK, although the same is not true for women (Zoratti 1998). African-Americans also have very high rates of obesity, metabolic syndrome and type 2 diabetes, and, in contrast to Afro-Caribbean populations in the UK, have high rates of coronary heart disease compared to the general US population (Hall *et al.* 2003). In addition, people of African origin living in the USA, Caribbean and UK have high levels of blood pressure (Cooper *et al.* 1997). One study in the UK, for example, found that 31% of Afro-Caribbean men and 34% of Afro-Caribbean women had hypertension, compared to 19% of white men and 13% of white women (Lane *et al.* 2002). Rural African populations have relatively low levels of hypertension (Cooper *et al.* 1997).

In most cases, then, these populations have a shared propensity, for whatever reason, to type 2 diabetes and cardiovascular disease when exposed to a western environment. Many also have high rates of obesity, particularly abdominal obesity. Below, I review suggestions put forward to explain these tendencies. Clearly, there are differences in the disease profiles of these populations, such as the lower rates of general obesity in South Asians and the high blood pressure without high rates of coronary heart disease in people of African origin living in the UK, and these differences have also been addressed to a limited extent.

The contribution of genes

Neel's thrifty genotype theory

In 1962 James Neel published the suggestion that the basic defect in diabetes mellitus was a quick insulin trigger in response to hyperglycaemia (high levels of blood glucose) (Neel 1962). Neel considered that the quick insulin trigger was under genetic control and coined the term 'thrifty genotype'. Rapid insulin release in response to a rise in blood glucose was hypothesised to lead to the rapid utilisation of glucose, minimising its loss in urine. Neel proposed that the thrifty genotype would have been selected for under conditions of 'feast or famine', because of a heightened need to make good use of glucose available during times of plenty, to see people through times when food was scarce. He believed that hunter–gatherers during the Palaeolithic would have experienced cycling of food availability, with periods of both abundance and scarcity. Neel considered that, in a western environment, a thrifty genotype would make people vulnerable to developing high levels of obesity and diabetes. Neel did not suggest any difference between populations in the prevalence of the thrifty genotype, but rather that because of the evolutionary background of the species, people in general would be vulnerable to the overproduction of insulin in the western environment, with some between individual variation in genetic vulnerability to diabetes. Thus, Neel should be credited with a very early exposition of the idea of the mismatch between 'Stone Age' genes and the modern western lifestyle.

Understanding of glucose and insulin metabolism and of the pathologies underlying diabetes was at a rudimentary stage when Neel put forward the thrifty genotype concept, and it later became clear that the specific physiological mechanisms he invoked were not plausible. As a result, he attempted to describe alternative pathways (Neel 1982), but he and colleagues later wrote that these modifications to the original idea were not 'intellectually satisfactory' (Neel *et al.* 1998). They also wrote that 'the term "thrifty genotype" has served its purpose, overtaken by the growing complexity of modern genetic medicine' (Neel *et al.* 1998, p. 60). They preferred to conceptualise obesity, type 2 diabetes and essential hypertension as resulting from previously adaptive multifactorial genotypes that had evolved to work together in an integrated manner in the Palaeolithic environment, but were no longer adaptive in a western environment, where an overabundant diet and sedentary behaviour act together to disrupt this integrated functioning. Neel *et al.* (1998) noted 'the suspicion that there might be a particular predisposition to the disease in some tribal groups', but suggested that there is no evidence for a simple 'ethnic predisposition'.

Modifications of the thrifty genotype theory

While Neel did not originally suggest that particular populations may have more 'thrifty genotypes' than others, and in his later work did not pursue the idea that some populations may be more vulnerable to developing high levels of type 2 diabetes on acquisition of a western lifestyle, this is the main direction in which others took the thrifty genotype concept. They were inspired by the underlying logic of the thrifty genotype theory, and suggested refinements aimed at explaining the very high rates of type 2 diabetes in particular populations. Subsequently, some commentators have come to think that these ideas are seriously flawed. I examine them here in some detail.

As discussed in Chapter 2, Palaeolithic hunter–gatherers were, for the most part, unlikely to have experienced severe cycles of famine and fasting (as acknowledged by Neel in his late discussions, e.g. Neel *et al.* 1998). Proponents of a modified thrifty genotype concept suggest instead that there may have been other populations that did have such experiences and that these

Fig. 4.2. The Arrival of the Māoris in New Zealand, by Charles F. Goldie and Louis John Steele 1899. Gift of the late George and Helen Boyd to the Auckland Art Gallery Toi O Tomaki. Reproduced with permission.

populations were therefore under selective pressure to develop thrifty genes. For example, McGarvey *et al.* (1989) suggested that selection for a thrifty genotype would have been stronger in temperate and semitropical agricultural systems, which are characterised by marked seasonality of food availability, susceptibility to drought and unpredictable rainfall and frosts.

More specifically, it was argued that the susceptibility of Polynesian populations to type 2 diabetes may result from a high prevalence of a thrifty genotype in these populations (Baker 1984). The Polynesian islands were populated by people who travelled by sea from Melanesia, starting about 3000 years ago. It has frequently been suggested that the high levels of type 2 diabetes seen in Polynesians arose as a result of the genetic consequences of selection exerted by their passage across large stretches of ocean between Melanesia and Polynesia, against prevailing winds and currents (Bindon and Baker 1997). It was suggested that, during these voyages, there must have been severe food shortages, and that only those who were efficient at storing calories, because they had a thrifty genotype, would have survived the longer voyages (Baker 1984). The physiological mechanism invoked was a large insulin response to carbohydrate intake, which could hasten the onset of type 2 diabetes in the western context (Bindon and Baker 1997). Furthermore, it was suggested that layers of fat would have provided important insulation during cold nights at sea, reducing energetic requirements for thermoregulation (Baker 1984). Baker also argued that, once the colonisers reached land, periodic starvation would have been a problem on the smaller islands, as a result of destruction wrought by periodic typhoons.

The idea of the devastating sea-voyage of the founding Polynesians is illustrated in a picture painted by Louis Steele and Charles Goldie in 1899, depicting the arrival of Māoris in New Zealand (Blackley 2001; Prentice 2001) (Fig. 4.2). However, shortly after the painting was shown in New Zealand, there were reports that Māoris who saw it were indignant at the manner in which it represented the arrival of their ancestors on the islands. Blackley gives the following quotation from a 1934 edition of the *Auckland Star* about the Māori response to this and similar work: 'Far from being appreciative, they always regard them with dubious feelings and disdain. To them they are mere creations of the pakeha [non-indigenous New Zealanders] mind and not consistent with the traditional records of the matters represented'. The complaint was that the painting diminished Polynesian maritime prowess (Blackley 2001, and that there was no oral history record of such suffering, but rather that their forebears were able to travel at sea indefinitely by catching rainwater and eating fish and turtles (Blackley 2001; Prentice 2001; Prentice *et al.*, 2005)). This may be a case where eurocentric assumptions have been applied to the development of a scientific hypothesis (Paradies *et al.* 2007). At the very least, the debate demonstrates that the attractive ideas of Prior and Baker and others about the experiences of the founding Polynesians are based partly on assumption.

The thrifty genotype idea was also applied to the Micronesian population of Nauru by Zimmet *et al.* (1977) who, invoking Neel's 1962 paper, suggested that the population might have a 'diabetic genotype'. The population of Nauru was also founded by people who undertook canoe voyages, which would have lasted several weeks. Building on this suggestion, Diamond (2003) noted that Nauruans also experienced extreme starvation and mortality during the Second World War, when the island was occupied by the Japanese military and half the population died as a result of low food rations, forced labour and deportation. Diamond suggests that this would also have resulted in selection for those with thrifty genotypes, who would have been more likely to survive such conditions. It is worth noting, however, that average daily dietary intake on Nauru was around 6100 calories in the 1970s, approximately double that of standard western intakes, and levels of physical activity had decreased, suggesting that, even without a particular vulnerability to insulin resistance, this population would be expected to have a high rate of type 2 diabetes (Zimmet *et al.* 1977).

One explanation offered for the very high rates of type 2 diabetes in the Pima of Arizona and other groups of Native Americans was that Native Americans had a genetic predisposition to 'New World Syndrome', a term which was used to describe the high prevalence of obesity, type 2 diabetes and cholesterol gallstones (Weiss *et al.* 1984). It was suggested that an improved ability to store fat was a central part of this genotype. This could, it was suggested, be part of a genetic tendency to be selectively insulin resistant, with insulin resistance in the muscles acting to ensure high levels of serum glucose, but allowing maintenance of the effects of insulin in promoting fat storage (Knowler *et al.* 1983). A line of evidence adduced in favour of this idea is that in groups where there had been mixing with non-Native American populations, individuals with 'full inheritance' had a higher prevalence of diabetes than those of mixed ancestry. For example, it has been shown that non-diabetic Pima had more European admixture than diabetic Pima (Williams *et al.* 2000). However, there may be lifestyle as well as genetic differences between full inheritance and admixed individuals and groups, and such differences are not adequately controlled for in this or similar analyses (Paradies *et al.* 2007).

Szathmáry (1994) suggested that a different kind of thrifty genotype may be prevalent among Native Americans. She noted that the original ancestors of modern Native Americans lived in the Arctic and Subarctic, where meat and fat are the only available foods for most of the year. She suggested that high protein and fat intake, together with carbohydrate restriction in a low temperature environment in which heavy physical activity was necessary, requiring the use of glucose as a source of energy, would lead to a selective advantage in those best able to convert non-carbohydrate sources into glucose. She notes that populations with high frequencies of genes that enhance the production of glucose would be particularly vulnerable to type 2 diabetes on adoption of the carbohydrate-rich western diet. Not dissimilar to Szathmáry's proposal is the Northern Hunting Adaptation hypothesis. Ritenbaugh and Goodby (1989) suggested that, for the same reasons, glucose sparing, where the breakdown of glucose is slowed and tissues other than the brain make use of fats as an alternative energy source, would have been beneficial in the Arctic environment. They expect that Native American descendants of the first North Americans should carry genetic adaptations to enhance glucose sparing.

Yet another similar suggestion has been made in relation to Australian Aborigines, who would have had periodic 'feasts' provided by kills of large animals, which would have had high protein but low fat content. Again, enhanced production of glucose by the liver would be an efficient system for converting excess dietary protein into glucose and fat as readily available forms of energy (O'Dea and Piers 2002), but in the presence of a western lifestyle would lead to an increased risk of type 2 diabetes.

While within the US ideas about thrifty genes have focused on Native Americans, and within Australia they have focused on native Australians, within the UK the group considered to be genetically susceptible to type 2 diabetes is South Asian immigrants. In the first studies of increased risk of type 2 diabetes and coronary heart disease in this group, the focus was strongly on the notion of a genetic vulnerability in South Asians (McKeigue *et al.* 1991). In 1995 an editorial in *The Lancet* provided a list of 'genetically acquired risk factors that are potentiated and supplemented by westernization' considered to be found in people of South Asian descent (Williams 1995). These included low function in the cells in the pancreas that secrete insulin, high levels of insulin resistance, and high plasma insulin levels. While mention is made of the thrifty genotype theory in this context (e.g. Abate *et al.* 2003), there has been little attempt to elaborate a reason why people in South Asia should have a thrifty genotype.

As westernisation and socio-economic development have proceeded, the number of populations with higher rates of type 2 diabetes than western populations of European origin has increased and the logic of suggesting that these populations each underwent a separate process of natural selection resulting in their carrying thrifty genotypes was questioned (Allen and Cheer 1996; Pollard 1997). One solution was to turn the concept around and consider whether a thrifty genotype might be the normal ancestral state in humans, with the exception worldwide being the 'non-thrifty genotype' (Allen and Cheer 1996).

Allen and Cheer suggested that all humans originally shared a thrifty genotype, but that special selection pressures encountered in Europe led to the evolution of a genotype that conferred resistance to diabetes in people of European ancestry. They note that famine was common in Europe until relatively recently, so the explanation cannot lie in an experience of a consistently better food supply. Instead, they suggest that it is the European population's use of milk and milk products that may have led to selective pressure against a thrifty genotype. They cite evidence that dairy products appear to have been used in Europe from around 6000 to 7000 years ago. Milk has a high sugar content compared to the foods available to hunter–gatherers or to most subsistence agriculturalists. It would have been a particularly valuable food for Europeans because it is a source of calcium in an area where levels of sunlight are relatively low, leading to vulnerability to calcium deficiency[1]. Allen and Cheer proposed that because of this plentiful source of serum glucose, Europeans were under selective pressure to be more sensitive to the effects of insulin, allowing the rapid clearance of glucose from the blood, exerting appropriate control over the use and storage of energy and avoiding the development of type 2 diabetes. We know that the use of milk imposed selective pressures that affected the frequency of different variants of the lactase gene (see p. 15), adding plausibility to Allen and Cheer's hypothesis. They suggest several ways in which the hypothesis might be tested, including investigations of the prevalence of type 2 diabetes in non-European populations with a long history of dairying (see p. 15) as they become westernised.

Others have proposed similar ideas without reference to Allen and Cheer. For example, one paper suggests that the use of honey in ancient European civilisation might have imposed selective pressure to increase insulin sensitivity (Baschetti 1998). Alternatively, Diamond's (2003) view is that, with a gradual increase in food supplies between the fifteenth and eighteenth centuries, Europeans experienced a slow rise in the prevalence of type 2 diabetes. He identifies J. S. Bach as one of the possible victims of what he calls a 'cryptic epidemic'. The implication from his general line of argument is that he considers that Europeans are now less vulnerable to type 2 diabetes because of the resulting selective pressure, which favoured those carrying relatively insulin sensitive genotypes. However, there is no evidence of a general increase in obesity in Europe at this time and the timescale is short for the effects proposed (Allen and Cheer 1996; Prentice *et al.* 2005).

The most important point to emerge from these papers is the suggestion that the 'thrifty genotype' may be ancestral, as originally proposed by Neel, and that 'non-thrifty genotypes' may have evolved subsequently in some populations. The suggestions about why non-thrifty genotypes might have evolved are amenable to further research. In summary, it makes less sense to expect derived 'thrifty' genes in non-European populations than it does to expect non-thrifty genes in people of European descent or in other descendants of populations with long-standing use of milk, or possibly other sources of glucose.

Attempts to identify genes conferring population-level susceptibility to type 2 diabetes

With the increased pace of research on genetic susceptibility to disease, there have been attempts to identify genes that confer susceptibility to type 2 diabetes. Given continuing fascination with the thrifty genotype hypothesis, despite the critiques noted above, some researchers have framed their search within the thrifty genotype paradigm.

One approach has been to start with an obvious candidate thrifty gene, such as the insulin gene. However, when the nucleotide sequences of the insulin gene were determined in groups of Pima and Nauruans, they were found to be the same as previously published sequences from other populations (Raben *et al.* 1991). More recently, with advances in technology it has become possible for geneticists to search for genes that might cause increased susceptibility to type 2 diabetes in the Pima and other Native Americans. One technique is to use genome-wide linkage scans, which involve a search of the genome for areas that are linked to the disease. If an area of chromosome is found in which variation appears to be linked to diabetes, a gene causally related to the disease may be situated there. Using this method, a gene on chromosome 11, which is associated with increased BMI and early onset diabetes within the Pima, has been identified (Hanson *et al.* 1998). This finding has been replicated in studies of other populations, including US populations of European origin and Mexican origin (Atwood *et al.* 2002; Duggirala *et al.* 2003; Baier and Hanson 2004). However, there has been no suggestion that a version of this gene that explains their particular disease pattern is found at high frequency amongst the Pima.

Linkage studies subsequently suggested that there might be a susceptibility gene for type 2 diabetes on chromosome 2, and this gene was later identified as that coding for calpain-10, an enzyme thought to have an important role in signalling between cells (Marshall *et al.* 2005). Variation at this locus was shown to be associated with type 2 diabetes within a Mexican American population and in two European populations (Horikawa *et al.* 2000). However, this finding has not consistently been replicated (Barroso 2005) and a meta-analysis concluded that the main risk variant was associated with only a small increased risk of type 2 diabetes (Weedon *et al.* 2003). There is some evidence that it has been subject to selection and variants are found at significantly different frequencies in different populations. Thus Fullerton *et al.* (2002) surveyed the calpain-10 gene across 11 populations from 5 continents, and found evidence of a history of positive natural selection. Even so, while calpain-10 has some potential to be considered a thrifty gene, it is clearly not going to explain the marked vulnerability of some populations to type 2 diabetes.

A recent study has found evidence that migrant South Asians in the USA have a higher frequency of a variant of the gene (PC-1) K121Q than do USA whites. This gene appears to be involved in cellular insulin signalling and the variant which was more frequent in South Asians was associated with greater levels of insulin resistance (Abate *et al.* 2003). It has been suggested that this variant 'confers a biological advantage, such as that postulated in the "thrifty gene hypothesis" ' (Carulli *et al.* 2005). While this idea is undeveloped, the gene appears to be a possible candidate for explaining part of the elevated risk of type 2 diabetes and coronary heart disease risk in people of South Asian origin living in westernised environments.

While the search for a genetic underpinning of type 2 diabetes has made slow progress, major genes underlying the rare familial diabetes syndrome, maturity-onset diabetes-of-the-young (MODY) have more readily been identified. People with MODY are lean, with low levels of plasma insulin resulting from defects in insulin secretion (Sellers *et al.* 2002; Hegele and Bartlett 2003). One of the genes that causes this condition is HNF1A, which codes for hepatocyte nuclear factor (HNF)-1*a*. (HNF)-1*a* is involved in the activation of many genes, including the insulin gene. One variant of HNF1A appears to be specific to the Oji-Cree. Although this variant does not cause MODY, it seems to reduce the age of onset of type 2 diabetes in this population by about 7 years for each copy of the mutated gene carried (Sellers *et al.* 2002; Triggs-Raine *et al.* 2002; Hegele and Bartlett 2003). Obesity and insulin resistance are common in the Oji-Cree, and it appears that, under these circumstances, individuals who also carry the HNF1A mutation are particularly vulnerable to the rapid onset of type 2 diabetes because they do not secrete enough insulin to allow them to control their high glucose levels (Triggs-Raine *et al.* 2002). Hegele and Bartlett (2003) note that the Oji-Cree population currently suffering from an epidemic of type 2 diabetes are the descendants of the survivors of a life that was previously very hard (see above), and included winter starvation. They also note that it is not known whether the HNF1A mutation would aid survival in such conditions, and it is possibile that the gene might have reached a high frequency in this population as a result of genetic drift or founder effect (that is, in small isolated groups such as the Oji-Cree it is possible for a new mutation to become common in the population simply by chance processes). Interestingly, they do not invoke the thrifty gene concept. It is also important to note that the high-risk HNF1A variant is not found in the majority of Oji-Cree diabetics (Hegele *et al.* 2000).

Overall, a very few genes have been identified as being associated with the risk of type 2 diabetes, and they explain to no more than a minimal extent the susceptibility of particular populations to disease. Similarly, there has been little success in identifying genes that might account for within population differences in vulnerability (Barroso 2005).

Genes for hypertension in populations of African descent?

A notion that shares many of the characteristics of the most common version of the thrifty genotype hypothesis is the 'slavery hypertension hypothesis'. This theory, put forward to explain the high rates of hypertension in African Americans, suggests that they carry genes that predispose to high blood pressure because of the experiences of their ancestors who were forcibly shipped, in terrible conditions, from Africa to the Americas. The most recent version of this hypothesis is that, during the trans-Atlantic voyage, high rates of mortality due to heat, vomiting, diarrhoea and general dehydration would have meant that the survivors who reached the Americas were those who were particularly good at conserving salt and water (Wilson and Grim 1991). Their descendants would then be more vulnerable than other populations to the development of high blood pressure when exposed to risk factors for high blood pressure. Such a mechanism would be likely to be effected through adaptation of the renin–angiotensin system, which plays the major role in controlling salt retention and regulation of the volume of blood. The hypothesis has been controversial. Both the historical evidence for the mortality associated with the voyage from Africa to the New World and the plausibility of the physiological mechanisms invoked have been strongly contested (Kaufman and Hall 2003). Furthermore, no publications have so far linked any particular genetic variation with the slave hypothesis. Nevertheless, the theory is reiterated consistently in the epidemiological literature (Kaufman and Hall 2003).

Another explanation posits a genetic predisposition to hypertension in those with African ancestry (not just the descendants of New World slaves), and indeed in all heat-adapted populations (Gleibermann 1973; Young *et al.* 2005). This hypothesis has the advantage of simplicity, proposing that those with ancestry in hot environments are more susceptible to hypertension than those with ancestry in cold environments. Here, the emphasis is not on race or ethnicity or the particular history of a particular population, but on the selective pressures imposed by climate. The authors note that humans have an unmatched capacity to sweat, which can lead to the loss of large amounts of salt and water. Considerable sweating, combined with low salt availability in tropical climates, imposes pressures to conserve salt in the body. Furthermore, losses in blood volume resulting from daytime sweating may lead to a need for increased contractility of the heart and arteries in order to maintain blood pressure. Enhanced salt retention and contractility of the heart and arteries may lead to an increased susceptibility to high blood pressure when exposed to lifestyle risk factors. Young *et al.* note that, as populations of *Homo sapiens* expanded out of Africa from about 100 000 years ago, the need for heat dissipation, and thus the need to conserve salt, was lost. Thus cold-adapted populations have less of a tendency to retain salt and less cardiovascular reactivity. Thus, it is claimed that populations with ancestry in hot areas retain ancestral genes that have been lost in cold-adapted populations such as Europeans. In support of their ideas they note that variation at two genes that influence blood pressure (AGT (angiotensin) and CYP3AG) also influences the retention of salt, and that for both genes the variant that increases salt retention is found more commonly near the Equator.

Young *et al.* (2005) expect that people descended from populations with a long history of living in a tropical climate will be more vulnerable to hypertension on the adoption of a western lifestyle. This theory has the strength of being based on climate, which is known to exert strong selective pressures on human biology (for example, on skin colour and on body shape). However, many researchers are convinced that other explanations are enough

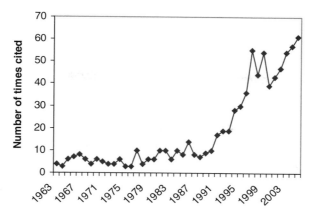

Fig. 4.3. Numbers of citations of Neel's original thrifty genotype paper (Neel 1962) by year. Figures from the ISI Web of Knowledge Service for UK Education.

to account for the high rates of hypertension in African Americans, as outlined below (pp. 72–73).

Why have genetic explanations proved so appealing?

The fact that interest in the thrifty genotype concept (mostly in its population-specific sense) has increased rather than receded in recent years is illustrated in Fig. 4.3, which shows the increasing number of citations of Neel's original paper. Some recent citations have been by those critiquing the predominant application of the thrifty genotype hypothesis, but these papers are in a considerable minority.

Many authors have criticised the tendency to assume that differences identified between different racial or ethnic groups are likely to reflect genetic differences between those groups (Braun 2002; Ellison and Rees Jones 2002). One criticism is that an emphasis on genetic explanations resurrects the notion of biological difference rooted in genetics, ignoring the long-standing discrediting of race as a biological category (Lewontin 1982). Fee (2006) notes that 'What is most striking about "the thrifty genotype" is how a rather unclear scientific hypothesis was transformed into a clearcut racializing account that is now a popular and free-floating "explanation" for the high incidence of diabetes in Aboriginal peoples'. She dissects Neel's original papers, and is one of very few authors to note that Neel did not link the thrifty genotype to particular populations, but that his writing has been taken by others to do so. She suggests that an overemphasis on 'race' and blindness to socio-economic history and culture is widespread. She notes that in the widespread concern about the obesity epidemic, one way for society to control fear of a disease is to represent it as afflicting someone else, just as HIV/AIDS is often seen either as a 'gay' or 'African' disease. The implication is that we need to be wary of assuming that differences between populations or between socioculturally constructed races are likely to be caused by differences at the genetic level.

A more quantitative approach to the issue of genetic vulnerability to these multifactorial diseases would be very helpful, and more biologically plausible than one based on ethnic or racial groups. This type of approach is inherent in the notion that heat-adapted populations may be particularly susceptible to hypertension, described above (Young *et al.* 2005). Here, variation is posited to be related to features of the physical environment rather than confined to a socially defined ethnic or racial group.

The influence of the environment experienced in early life

Developmental origins of adult disease

A surge of interest in early life influences on the development of disease in later life began with the publication of studies showing that areas of Britain with the highest rates of neonatal mortality early in the twentieth century tended to have the highest rates of coronary heart disease later in the century (Barker and Osmond 1986). David Barker's group then found evidence of links between birthweight, as well as other measurements at birth, and later risk of cardiovascular disease, with small babies having higher risk (Barker 1994).

Many more studies have now been published, to the extent that it is possible to undertake meta-analyses and systematic reviews, summarising the findings of a large number of studies looking at specific links between birthweight and various risk factors. These studies show an inverse association between birthweight and central or abdominal obesity, so that smaller babies have the more harmful type of fat patterning (Oken and Gillman 2003). A systematic review of studies of links between low birthweight and later risk of type 2 diabetes found that 13 out of 16 reported that babies with lower birthweight had a greater risk of type 2 diabetes (Newsome *et al.* 2003). Other reviews have shown that the large number of studies so far conducted also provide evidence of an association of low birth weight with high blood pressure, as well as with high total cholesterol levels (Huxley *et al.* 2002; Huxley *et al.* 2004). These studies have generally focused on babies who were born small for their gestational age, but there is also evidence that babies born early (and therefore small) have a relatively high risk of developing type 2 diabetes in later life (Harder *et al.* 2007).

The combination of lower birthweight and higher BMI in adult life is most strongly associated with later disease risk. Thus, people who are born small, who have a more central type of fat patterning throughout life, and then put on weight and become heavy, are at greatest risk of type 2 diabetes (Rich-Edwards *et al.* 1999).

However, there is good evidence that studies that find these associations are more likely to get published than studies that do not find them, and therefore it seems likely that the associations are perhaps not as strong as initially claimed (Huxley *et al.* 2004). A further caveat is that some studies have identified a U-shaped relationship between birthweight and some of these outcomes, indicating that the largest babies also have an increased risk of disease (Harder *et al.* 2007). This phenomenon may be explained by the effects of a maternal tendency to insulin resistance, which increases birthweight in offspring and also appears to increase the offspring's risk of insulin resistance later in life (Drake and Walker 2004).

More recently there has also been a recognition that birthweight, or other characteristics at birth, are not the critical factors here, but happen to be the most easily available markers of early development. Further, it is now clear that it is not only development *in utero* which has implications for later disease risk and that postnatal development is also important. For example, a recent study of people born in Helsinki found that, on average, adults who had been admitted to hospital with coronary heart disease or who had died from coronary heart disease had been relatively small at birth and thin at 2 years of age and had afterwards put on weight rapidly (Barker *et al.* 2005). Unfortunately, it is difficult to separate the effects of low birth weight and rapid infant growth, since small babies tend to grow relatively fast (Simmons 2005).

It has been suggested that these associations arise because of changes that an individual makes to his or her developmental trajectory in response to nutritional and other restrictions on growth experienced in early life. Given that the neonatal and weaning periods are high risk times for humans, it may be that the selective pressures driving these adjustments derive from the need to get individuals through these early stages of life. For example, being a small baby with large central fat reserves (abdominal fat is mobilised more easily than subcutaneous fat) is likely to increase chances of survival during the first year or so of life. But there may also be longer-term benefits. Being on a high growth trajectory is advantageous if the conditions make it possible, because a fast-growing individual will reach reproductive maturity more quickly, will be a large adult and will increase his or her reproductive success by having greater ability to invest in reproduction (Hill and Kaplan 1999). Being on a slower growth trajectory will, however, be advantageous in an environment where there are constraints, such as limited nutritional supplies, which make it necessary to compromise in order to reach reproductive maturity successfully.

In these circumstances other adjustments are also likely. For example, it might be beneficial to have a reduced kidney size with a reduced number of nephrons (tubules where waste is removed from the blood), because the kidney is a major consumer of energy and demands a high blood flow. This leads to increased vascular resistance, which is then expected to predispose the individual to hypertension, but that is a problem which would not arise in the environment in which humans evolved (Gluckman and Hanson 2005, pp. 135–136). Gluckman and Hanson describe the adjustments that appear to be made in response to expectations of what the environment will be like in future life (based on signals received *in utero,* or perhaps in infancy and early childhood) as predictive adaptive responses. The suggestion is that the foetus is assessing the environment and making long-term adaptive adjustments to that environment.

On the other hand, developmental changes that impair health in later life may result simply from disruptive effects on normal development that cause long-term dysfunction in affected organs or tissues. In the case of type 2 diabetes, non-adaptive associations between markers of poor foetal growth and later disease may arise because the foetus receives insufficient resources, resulting in impaired development of the pancreatic beta-cells that secrete insulin. In those who develop insulin resistance in later life, as is common in western environments, an impaired ability to secrete insulin will hasten progression to type 2 diabetes (Weir *et al.* 2001).

Thus a number of explanations for the existence of developmental effects on later disease risk in western or westernising environments have been proposed and this debate is ongoing. However, there is now a reasonable consensus that such effects exist, so that restrictions on growth and development in early life increase vulnerability to some western diseases in later life.

Implications for rapidly westernising populations

This is a burgeoning field with much remaining to be discovered, but at its simplest the implication is that those born into relatively poor environments and later exposed to a western environment are particularly vulnerable to a variety of 'western' diseases. If the environment remains poor, the individual is not affected. It has been shown in The Gambia that small babies maintain excellent cardiovascular health into adulthood, with a complete absence of the metabolic syndrome, so long as they retain their 'lean, fit and frugal' lifestyle (Moore *et al.* 2001). But, if the individual adopts a more western lifestyle, he or she is particularly prone to the development of abdominal obesity, insulin resistance, high blood pressure, high cholesterol levels and thus type 2 diabetes and cardiovascular disease. Such individuals have been described as having a 'thrifty phenotype', marking the fact that this argument offers an alternative to the 'thrifty genotype' explanation for high risk of type 2 diabetes and cardiovascular disease (Hales and Barker 1992; Hales and Barker 2001). Of course, genes that allow for this plasticity must underlie the thrifty phenotype, but the important point is that the environment experienced early in life is key in determining the eventual phenotype and risk of disease.

A number of researchers are now pursuing these ideas in India in particular. It makes good sense to do this work in India, since birthweights in South Asia are amongst the lowest in the world (Kramer 1987), socio-economic development is rapid, and, as we have seen, rates of type 2 diabetes and coronary heart disease in South Asians living in the west are very high. Studies have shown that, although small in comparison to babies born in the UK, Indian babies have a relatively high proportion of central body fat compared to British babies (Yajnik *et al.* 2002). Further, birthweight is inversely associated with insulin resistance in Indian children and it is those children who were the smallest at birth and the heaviest at the time of testing who have the highest levels of insulin resistance (Yajnik 2000).

Thus, what is often known as 'the Barker hypothesis' offers an alternative explanation for the high rates of type 2 diabetes seen in many other rapidly westernising populations around the world. Nauru, for example, experienced a very rapid increase in wealth, following the exploitation of its extensive phosphate reserves. Similarly, many Native Americans and Australian Aborigines have abandoned their traditional modes of subsistence and have settled and adopted diets that are much higher in sugar and energy in general than those of previous generations. For example, the traditional Pima diet of grains, squash, melons and legumes, supplemented by gathered desert plants, has become increasingly westernised (Smith 1994). The current Pima diet provides energy and major nutrients in a proportion similar to the general US diet and popular components of the diet include a typical American-style breakfast of eggs, bacon or sausage and fried potatoes, Mexican-style foods such as tacos and refried beans, hamburger and pork chops, as well as traditional Pima desserts such as cholla (cactus) bud stew (Smith *et al.* 1996).

Does this mean that one generation will be badly affected by the adoption of a western lifestyle, but that future generations will not suffer because they have received the 'correct' signals during gestation and early life, having been born into a westernised environment? The answer, unfortunately, is probably not, for two important reasons. Development *in utero* is affected by a number of factors, including maternal nutrition, maternal infection (particularly malaria) and cigarette smoking, all of which have important effects during gestation (Kramer 1987). Even if women have higher energy intakes during pregnancy, other dietary deficiencies and increased rates of smoking in women in many poor countries (Asia–Pacific Cohort Studies Collaboration 2005), together with high rates of infectious diseases, are likely to impose restrictions on the growth and development of babies (James 2002).

Second, the development of the foetus is also affected by whatever has previously determined the mother's size (James 2002). Smaller women have smaller babies (Kramer 1987). Kuzawa (2005) has suggested that this is likely to be partly the result of evolved mechanisms. Because the environment experienced during gestation may not accurately reflect the longer-term environment, for example, because of seasonal effects, he suggests that messages are passed down the maternal line from generation to generation to give the developing baby a longer-term measure of the external environment. This process creates what he calls 'intergenerational phenotypic inertia'. The implication of this long-term cueing of developmental processes *in utero* is that babies born to women who may themselves live in a western environment, but whose recent maternal ancestors lived in relatively poor environments, are likely to follow developmental pathways which make them vulnerable to the development of type 2 diabetes and cardiovascular disease in later life.

Unfortunately, even the slow washing out process implied by these ideas is likely to be impeded by the effects of insulin resistance and high levels of glucose in mothers during pregnancy, which also have intergenerational consequences, increasing the risk of type 2 diabetes in their offspring (Fall 2001; Drake and Walker 2004). For example, the offspring of Pima women who are diabetic during pregnancy have a particularly high prevalence of young-onset type 2 diabetes (Pettitt *et al.* 1988).

Reversing the question to consider why people of European ancestry have lower rates of type 2 diabetes and cardiovascular disease compared to other groups, as in the 'non-thrifty genotype' argument, is also useful. People of European ancestry have made the transition to a westernised lifestyle more slowly than many other groups around the world and thus the argument is that they are less likely to experience developmental influences, which prepare them for an environment different to the environment(s) they actually experience as adults.

The thrifty phenotype hypothesis offers a relatively parsimonious solution to the puzzle posed by the very high rates of type 2 diabetes and cardiovascular disease seen in so many rapidly westernising populations around the world. If we accept this proposition, the implications are worrying, because we can predict a huge increase in rates of type 2 diabetes as populations increasingly undergo westernisation, and this will be a continuing phenomenon, not limited to a single generation.

Reconciling the two explanations for 'thriftiness'?

Although the thrifty genotype and thrifty phenotype approaches to explaining high rates of type 2 diabetes and cardiovascular disease in some populations may seem very different, there are possibilities for reconciling and integrating them (Hales and Barker 1992). One development of possible significance is that two genes have been identified that affect both foetal growth and risk of type 2 diabetes (Ong and Dunger 2000; Ong and Dunger 2004). Similarly, the foetal insulin hypothesis suggests that genetically determined insulin resistance results in impaired insulin-mediated growth in the foetus as well as insulin resistance in adult life (Hattersley and Tooke 1999). However, there has thus far been no answer to the question of whether the populations with the highest rates of type 2 diabetes also carry particular variants of these genes.

Another fascinating line of enquiry suggests that transgenerational effects on risk of metabolic syndrome may be transmitted via epigenetic mechanisms (Gallou-Kabani and Junien 2005). Here, the environment experienced by an individual, particularly in early life, is thought to affect the expression of his or her genes in ways that are heritable by future generations. It seems likely that future work to understand these mechanisms will be fruitful in establishing inherited and early life effects on vulnerability to type 2 diabetes and cardiovascular disease.

Other explanations

While the thrifty genotype and thrifty phenotype hypotheses have attracted a great deal of interest and stimulated much important research, there are other possible explanations for between population variation in rates of obesity, type 2 diabetes and cardiovascular disease.

Exposure to infectious disease

We saw in the previous chapter that exposure to certain, usually chronic, infections may be a risk factor for later development of some non-communicable diseases, including perhaps type 2 diabetes and cardiovascular disease. Infectious diseases are more prevalent in poorer populations around the world, and are also more common in native populations in North America, Australia and New Zealand than in the general population (Gracey and Spargo 1987; Scragg *et al.* 1996; Holman *et al.* 2001). Thus this is another possible reason why some populations may be at increased risk of type 2 diabetes and cardiovascular disease when they take on a western lifestyle.

Racism

One of the strongest explanations currently offered for the high rates of blood pressure in African-Americans focuses on their experiences of racism in daily life. There is good evidence that African-Americans suffer significantly from racist practices and attitudes (Williams 1999), and it has been shown that perceived racism is positively related to blood pressure during everyday life in African-American men and women (Steffen *et al.* 2003). The link is thought to be mediated by the effects of anger, including the inhibition of anger, on blood pressure. There has been less of a focus on racism as a risk factor for poor health amongst other ethnic groups. However, a recent study showed that, in New Zealand, Māori were more likely to report experiences of racial dis- crimination and that this discrimination contributed to Māori ill health (Harris *et al.* 2006).

Poverty

When the work by Barker and his team identifying links between regional variation in infant mortality rates in the early twentieth century and rates of coronary heart disease in the late twentieth century was first published, other researchers pointed out that the association could have arisen as a result of regional variation in poverty (Ben-Shlomo and Davey Smith 1991). Infant mortality would have been highest in the poorest areas and, by the end of the twentieth century, it was the poorest members of society who had the highest rates of coronary heart disease. Researchers working on developmental origins of adult disease have refuted this argument by showing the resilience of associations between markers of development and later cardiovascular health across different populations and by work iden- tifying mechanisms by which developmental influences may increase risk of coronary heart disease in later life. However, the effects of poverty and low socio-economic status are very real and should not be forgotten in attempts to explain between population variation in obesity, type 2 diabetes and cardiovascular disease.

In general, socio-economic inequalities are likely to be fundamental causes of inequalities in health between ethnic groups (Nazroo 2003). For example, in the United States, African-Americans have less income and education than whites on average, both of which are related to increased cardiovascular morbidity and mortality (Williams and Collins 1995). The fact that in the USA, African-Americans have a higher risk of dying from coronary heart disease than the general population is thought to be accounted for by effects of socio-economic status. For example, in one study the relative risk of death from coronary heart disease was no different for blacks and whites when people of the same socio-economic status were compared (Keil *et al.* 1992). It has been suggested that the higher risk relative to the general population of coronary heart disease in African-Americans compared with people of African/Caribbean origin living in the UK might be partly explained by lower socio-economic status and poorer access to health care among African-Americans (Forouhi and Sattar 2006).

Socio-economic inequalities also underlie a great deal of the health disadvantage experienced by the native people of North America and Australia. For example, one study found that Aboriginal people living in the Six Nations Reservation in Canada had significantly lower levels of education and employment than people of European origin living nearby, and that over a third of the Aboriginal people lived in poverty. Rates of obesity and smoking were very high amongst the Aboriginal people, and were inversely related to income (Anand *et al.* 2001). In Australia, Aborigines are the poorest group in society (Tatz 2005).

Socio-economic status can affect health by many pathways. For example, those identified as operating in the USA include differential access to medical care, differences in health behaviours (most importantly, higher levels of smoking in those of lower socio-economic status), working conditions, environmental exposures, for example to hazardous waste sites, differences in employment rates and economic hardship and power differentials (Williams and Collins 1995). Explanations for the very high rates of obesity-related diseases in the populations considered here that focus mainly on their relative poverty acting via any of these pathways are not as novel or exciting as the thrifty genotype and thrifty phenotype hypotheses and have not received the attention they deserve. It is easier to attempt to identify genetic vulnerability in some groups than it is to tackle problems associated with poverty.

Summary

In the 1970s there were a few populations with rates of obesity, type 2 diabetes and cardiovascular disease that seemed markedly high compared to people of European descent living in affluent countries. Explanations centred initially around the thrifty genotype concept, but most applications of this concept were and are flawed. Several authors have sounded a cautionary note about the appeal of racialised and Eurocentric versions of the thrifty genotype concept. It makes more sense to expect a 'non-thrifty' genotype in Europeans to explain their relatively low susceptibility to type 2 diabetes. Progress to identify genes that can account for between population differences in susceptibility to these diseases has been slow. This does not exclude the possibility that certain non-European populations are particularly genetically susceptible to type 2 diabetes because of their history but it is clear that there are good reasons to be much more circumspect of such expectations than many have been in the past.

The finding that people who grow slowly in early life have a greater risk of type 2 diabetes and cardiovascular disease in later life on exposure to a western environment offers a simpler explanation for what has happened in populations that have experienced rapid changes in lifestyle. This approach offers a great deal of promise, although, as with the thrifty genotype, we must be wary of being too keen to jump uncritically onto this bandwagon. In particular, we must not forget other more prosaic explanations for the ill health of these populations, which are often exposed to other risk factors such as high rates of infection, racism and poverty.

Subsequent to this initial debate, it became clear that the populations highlighted here were only the first to experience rates of type 2 diabetes and cardiovascular disease that are much higher than those seen in Europeans. As other populations around the world are affected by western socio-economic forces, they suffer in a similar manner. The future health profile of such westernising populations is considered further in Chapter 9.

END NOTES

[1] Sunlight stimulates the synthesis of vitamin D, which is needed to help calcium uptake.

5 *Reproductive cancers*

Breast cancer is one of the biggest causes of mortality in western societies. Other reproductive cancers are less well known but also common, including endometrial and ovarian cancers in women and prostate cancer in men. They are often cited as examples of 'western' cancers, since levels are much higher in affluent western populations than elsewhere. As with most other cancers, and non-communicable diseases in general (see Chapter 2), increased lifespan is one obvious explanation of high rates of these diseases in affluent western populations, but increased lifespan cannot explain all of the association between westernisation and reproductive cancers. So, why does an affluent western lifestyle bring an increased risk of reproductive cancers?

In this chapter I draw on the work of biological anthropologists and epidemiologists to answer this question. After summarising geographical and temporal trends in the incidence of reproductive cancers, I describe the main known risk factors for these diseases. I then go on to show that the affluent western lifestyle has increased exposure to all of these risk factors, focusing in particular on the evidence linking western life to increased exposure to endogenous gonadal hormones: oestrogen and progesterone secreted by the ovaries in women, and androgens secreted by the testes in men. Here I draw particularly on the pioneering work of Ellison (Ellison 1999).

What are reproductive cancers?

Cancerous tumours, including those of the reproductive system, result from the unregulated division of undifferentiated cells. If untreated, cancer eventually spreads from its initial location to other parts of the body, invading other tissues and disrupting their function so severely that, at this stage, it is usually fatal. The development of cancerous tumours is caused by mutations in genes involved in the control of cell division, cell death and DNA repair. Such mutations are most likely to develop in tissues that experience a lot of cell division, because every time the DNA of a cell is copied there is an opportunity for a carcinogenic mutation to arise. The epithelium of the breast, the uterine endometrium (the lining of the uterus) and the granulosa cells of

75

ovarian follicles all experience regular waves of cell division associated with the menstrual cycle (Ellison 1999), while the prostate experiences chronically high levels of cell division. This explains the fact that cancers of the breast, endometrium, ovaries and prostate are all relatively common.

Geographical and temporal trends in the incidence of reproductive cancers

Breast cancer is by far the most common cancer in women, accounting for 23% of all cancers in women worldwide. It has a relatively good prognosis compared to other cancers, at least in western countries, but nevertheless is the leading cause of cancer mortality in women worldwide, causing 14% of all female cancer deaths (Parkin *et al.* 2005). More than half of all cases occur in industrialised countries and incidence rates are high in most of the affluent areas of the world, except for Japan (Fig. 5.1) (Parkin and Ferñandez 2006). North America and the United Kingdom have the highest age-standardised incidence rates (McPherson *et al.* 2000; Parkin and Ferñandez 2006) and, while the incidence is lower in eastern Europe, South America, southern Africa and western Asia, breast cancer is still the most common cancer of women in these regions (Parkin and Ferñandez 2006). In contrast, low rates are found in most African and Asian populations. The high recorded incidence in more affluent parts of the world is partly due to the presence of screening programmes, which mean that most cases of breast cancer are identified, whereas in other parts of the world some may not be identified. In recent decades breast cancer incidence rates have risen worldwide, although since the mid 1990s there has been some sign of a levelling off of this trend in North America and Europe (Botha *et al.* 2003; Parkin and Ferñandez 2006).

Ovarian cancer is the sixth most common cancer and the seventh most common cause of death from cancer in women worldwide (Parkin *et al.* 2005). Again, incidence rates are highest in developed countries, with Japan having the lowest rate amongst developed countries (Fig. 5.1). Incidence rates have been slowly increasing in many western countries and in Japan (Parkin *et al.* 2005). Similarly, incidence rates for endometrial cancer are up to ten times higher in western industrialised countries than in Asia or rural Africa (Kaaks *et al.* 2002).

In 2002 prostate cancer was the second most common cancer in men worldwide, after lung cancer, accounting for 12% of cancers in men (Parkin *et al.* 2005). A particularly good prognosis means that it accounted for only around 6% of all cancer deaths in men in 2002. As with breast cancer, some areas of the world practise more screening than others, and this influences the

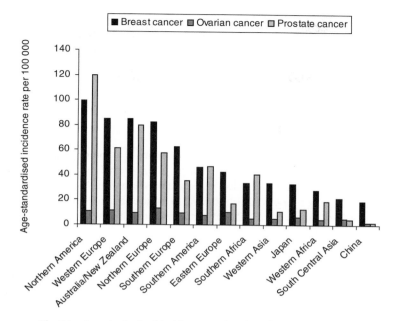

Fig. 5.1. Age-standardised incidence rate (number of new cases diagnosed) of breast cancer, ovarian cancer and prostate cancer by region for 2002. Data from Parkin *et al.* (2005).

detection rate and comparisons between countries. Thus, in the United States, where screening is common, prostate cancer is by far the most commonly diagnosed cancer in men (Parkin *et al.* 2005). Incidence is also high in northern and western Europe and in Australia and New Zealand, but much lower in Asia (Fig. 5.1). Within Asia it has been noted that the degree of westernisation correlates with the incidence of prostate cancer, with the highest rates in Japan, Singapore and Hong Kong (Hsing and Devesa 2001). Another striking feature of prostate cancer epidemiology is differences between ethnic groups in the same country, with US blacks suffering from particularly high rates (Hsing and Devesa 2001). In low-risk countries there was a large increase in the incidence of reported prostate cancer in the last quarter of the twentieth century, some, but not all, of which may be due to greater awareness of the disease (Parkin *et al.* 2005).

Testicular cancer, the other main reproductive cancer in men, is, in contrast, relatively rare, accounting for only 1–2% of all cancers in men and boys, although because it is often seen in younger men, it accounts for most cases of cancer in young men in developed countries (Huyghe *et al.* 2003).

The highest rates are observed in western Europe, northern Europe, Australia and New Zealand and North America and there is a low incidence in Asia, suggesting that, as with the other reproductive cancers, it is strongly associated with an affluent western lifestyle (Parkin *et al.* 2005). A rapid increase in incidence has been observed in most countries over the last 20–30 years, although mortality has declined because of the introduction of effective chemotherapy in the mid 1970s (Parkin *et al.* 2005).

Known risk factors for reproductive cancers

Genetic factors, including variation in the major susceptibility genes (BRCA-1 and BRCA-2) may account for up to 10% of breast cancer cases in developed countries (McPherson *et al.* 2000), and help explain why women with a family history of breast cancer are at increased risk. BRCA-1 and BRCA-2 also appear to play a role in affecting risk of prostate cancer (Kraft *et al.* 2005). Some ethnic groups, for example, Ashkenazi Jews, carry high frequencies of mutations in BRCA-1 and BRCA-2 that are associated with breast cancer (including early-onset breast cancer) and prostate cancer. These high frequencies, which would be difficult to explain in relation to the processes of natural selection, are thought to have arisen by founder effect, that is, by the action of random effects that can rapidly increase the frequencies of genes, even harmful genes, in small populations (Neuhausen 1999). However, the prevalence of harmful mutations in BRCA-1 and BRCA-2 is too low to explain much of the international variation in breast and prostate cancer risk (Parkin *et al.* 2005). A great deal of effort has gone into the attempt to identify other genes that may increase the risk of reproductive cancers, but there has been very little success, particularly in the identification of genes that could account for differences in cancer rates between populations (Platz and Giovannucci 2004). Thus the role of genetic variation in explaining between population variation in rates of reproductive cancers is likely to be small.

For all these cancers, geographical patterns of incidence as described above, changes in incidence rates following migration from low-risk to high-risk areas, and changes in incidence rates in populations experiencing industrial and socio-economic development associated with westernisation, strongly suggest that an affluent western lifestyle brings an elevated risk. For example, an early study of migrants from Japan to the United States showed that rates of breast cancer in the immigrant population reached similar rates to those of the host country within one or two generations (Dunn 1975). A more recent study of Chinese, Japanese and Filipino women living in the USA created an index of westernisation based on place of birth and migration history and showed that the risk of breast cancer increased with western-isation; length of exposure to a western lifestyle had a substantial impact on breast cancer risk for women born in China, Japan or the Philippines, while those born in the west had an even higher risk (Ziegler *et al.* 1993). Similarly, studies of prostate cancer risk suggest that environment is the main deter-minant of risk. Thus migrants from Japan to the USA have a much higher incidence than those still living in Japan, although populations of Asian origin (Japanese, Chinese and Korean) in the USA have lower rates than the general population of the USA (Parkin *et al.* 2005). Within Japan, increasing westernisation, especially of dietary habits, has been correlated with increasing rates of breast, ovarian and prostate cancer (Tominaga and Kur-oishi 1997).

An affluent western lifestyle is thought to increase the risk of reproductive cancers through several routes. The most important risk factor, at least for female reproductive cancers, is probably increased lifetime exposure to endogenous gonadal hormones. Features of reproductive life known to increase breast cancer risk, such as an early age at menarche, late age at menopause and lack of breastfeeding, are probably important because they are markers for high lifetime oestradiol and progesterone exposure. High levels of obesity and associated insulin resistance are also important risk factors for reproductive cancers, probably partly, but not entirely, because they are associated with increased exposure to endogenous gonadal hor-mones, particularly in relation to oestrogen levels in post-menopausal women. Having no children (nulliparity) or a late first birth, both features of affluent populations, are major risk factors for breast cancer. Much less well established are the possibilities that exposure to exogenous oestrogens and disruption of circadian rhythm are linked to increased rates of reproductive cancers. In this section I explore these risk factors in more detail, outlining the mechanisms by which they are thought to exert their effects on repro-ductive cancer risk.

Endogenous gonadal hormones

Women

The two main gonadal hormones in women are oestradiol and progesterone, produced by the ovaries. Oestradiol is the most important and potent form of oestrogen. The other forms of oestrogen, oestrone and oestriol, are mainly produced outside the ovaries by conversion from androgens, which are secreted mainly by the adrenal glands. Progesterone and oestradiol are pro-duced in significant quantities by the ovaries only during reproductive life, that is, after menarche and before the menopause. In post-menopausal women

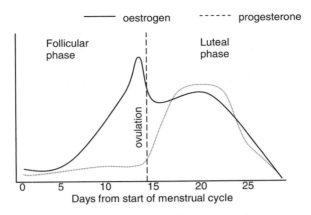

Fig. 5.2. Average patterns of variation in oestrogen and progesterone over a menstrual cycle.

most endogenous oestrogen is derived from conversion of androgens into oestrone and then oestradiol in adipose tissue. In post-menopausal women, unlike in pre-menopausal women, plasma oestrone levels are substantially greater than plasma oestradiol concentrations.

In women of reproductive age the level of secretion of progesterone and oestradiol varies across the menstrual cycle (Fig. 5.2). In the first half of the cycle, the follicular phase, when several follicles develop within the ovaries, the level of secretion of oestradiol gradually rises, while progesterone levels remain minimal. Oestradiol stimulates proliferation of the lining of the endometrium, leading to the development of an appropriate environment for implantation of an embryo, should conception occur later in the cycle. At ovulation, which occurs at mid-cycle, usually one dominant follicle produces an egg, that travels down the fallopian tube to the uterus. Once this has happened, the primary follicle changes form to become the corpus luteum, which then secretes large amounts of oestradiol and progesterone during the second part of the menstrual cycle, known as the luteal phase. These hormones sustain an environment that would allow implantation of an embryo in the endometrium, should conception have occurred. In the absence of a successful implantation, the corpus luteum degenerates, levels of oestradiol and progesterone drop and menstruation begins. This process is controlled by the secretion of hormones (called gonadotrophic hormones) from the pituitary gland, which is in turn controlled by the hypothalamus. The whole cycle usually takes around 28 days in western women.

In the blood, oestradiol is largely bound to the proteins sex hormone binding-globulin (SHBG) and, to a lesser extent, albumin. Bioavailable oestradiol, that is oestradiol that is available to target organs and tissues, is now usually considered to consist of the unbound fraction of oestradiol (free oestradiol) as well as that bound (loosely) to albumin. Thus the effects of oestradiol in the body depend on the level of oestradiol secreted by the ovaries, but also on the level of SHBG. Interpretation of studies of oestradiol is complicated by the fact that they sometimes measure total oestradiol, sometimes bioavailable oestradiol and sometimes free oestradiol.

Ovarian hormone levels play a key role in fecundity (the ability to conceive) and lower levels of oestradiol and progesterone during a menstrual cycle are associated with a lower probability of conception during that cycle (Lipson and Ellison 1996). Ovarian hormones, particularly oestradiol, also have wider effects across the body, playing an important role in pregnancy and influencing bone density, serum lipid profile, the skin and cognitive function (Mendelsohn and Karas 1999; Palacios 1999) (see discussion of the menopause in Chapter 6).

There is substantial experimental, epidemiological and clinical evidence that breast cancer risk is influenced by lifetime exposure to endogenous gonadal hormones in women (Bernstein and Ross 1993). The main focus of research in this area has been oestrogen with less of a consensus about the role of progesterone (Bernstein and Ross 1993; Muti 2004). The actions of oestrogen and progesterone in stimulating cell division in the breast, endometrium and ovaries provide a plausible mechanism linking increased lifetime exposure to increased cancer risk (Jasieńska and Thune 2001b). In the case of endometrial cancer, the action of unopposed oestrogen, that is high oestrogen levels in the absence of high progesterone levels, is thought to be most important (Henderson *et al.* 1982). This theory originated partly from observations that endometrial proliferation (cell division) takes place mostly during the follicular part of the menstrual cycle, when progesterone levels are low, and that women using medications containing oestrogens without progestins have increased endometrial cancer risk (Kaaks *et al.* 2002). It has also been hypothesised that exposure to high levels of oestrogen *in utero* is associated positively with breast cancer risk in later life (Ekbom 2006).

A woman's exposure to ovarian steroids during her reproductive years depends on the level of hormone secreted by the ovaries during a menstrual cycle, and on the number of menstrual cycles experienced during her lifetime. The number of menstrual cycles experienced depends on age at menarche, age at menopause and factors that prevent cycling between these ages, principally pregnancy and breastfeeding. During pregnancy, the production of ovarian steroids is high, but when a woman is breastfeeding intensively the ovaries are shut down, a state known as lactational amenorrhoea. Thus, as noted above, the well-established associations of early menarche, late menopause and limited breastfeeding with an increased risk of breast cancer are likely to be explained by the effect of each of these factors in increasing exposure to ovarian hormones over a lifetime, by extending the time during which the ovaries actively secrete hormones. Similarly, the removal of both ovaries before natural menopause has a significant impact on breast cancer risk and the earlier in life that the ovaries are removed, the greater the risk reduction (Trichopoulos *et al.* 1972). Similar associations have been observed for ovarian cancer, which is also thought to be strongly influenced by lifetime exposure to endogenous oestrogen (Zografos *et al.* 2004). In this case, of course, removal of the ovaries eliminates the risk of cancer.

Men

The testes secrete several androgens, including testosterone, dihydrotestosterone and androstenedione, but testosterone is found at much higher levels than the other two and is usually considered to be the most significant androgen. Much (40%–65%) circulating testosterone is bound to SHBG, with testosterone having an even higher affinity for SHBG than oestradiol. A similar proportion of testosterone is bound to albumin, leaving around 2% of testosterone 'free'. As with oestradiol, different studies measure different fractions of testosterone, sometimes total testosterone, sometimes (as in salivary studies) free testosterone, and sometimes bioavailable testosterone, now usually defined as free plus albumin-bound testosterone.

Testosterone has many roles within the body, including being responsible for the development of male sex-specific characteristics at puberty. Although it is necessary for the production of sperm, variation in testosterone levels does not have much effect on this process (Ellison 2001). Testosterone is an anabolic steroid, stimulating the growth of muscle, increasing the size and strength of bones and also stimulating the breakdown of fat (Herbst and Bhasin 2004). It plays an important role in determining the allocation of energy to reproductive effort as opposed to survival in men (Ellison 2001). Higher levels of testosterone lead to greater muscle mass (Herbst and Bhasin 2004) and in many animals, including humans, greater muscle mass is associated with greater success in competition with other males and increased attractiveness to women, both of which are likely to increase reproductive success (Bribiescas 2001a). Higher levels of testosterone are also associated with greater levels of aggression or assertiveness and enhanced libido in men, although the causality of these associations is unclear (Ellison 2001). However, maintaining high testosterone levels could have fitness costs because it appears to cause suppression of the immune system and to increase susceptibility to parasite infection (Muehlenbein and Bribiescas 2005).

There are good reasons to believe that androgens play a central role in the development of prostate cancer. For example, the growth and maintenance of the prostate are dependent on androgens, and prostate cancer regresses following anti-androgen therapy (Hsing 1996). Furthermore, higher levels of serum testosterone and lower levels of SHBG have been shown to be associated with an increased risk of prostate cancer (Gann *et al.* 1996; Shaneyfelt *et al.* 2000). While it has been suggested that exposures to androgens early in life (including during foetal development) may be critical, there is insufficient evidence to assess whether androgen exposure has different effects on prostate cancer risk across the lifecourse (Hsing 1996). Oestrogen exposure in early life, including that resulting *in utero* from high maternal oestrogen levels, may also play a role in determining prostate and testicular cancer risk (Sharpe and Skakkebaek 1993; Ekbom 2006).

Obesity

Obesity is a strong risk factor for endometrial cancer and for post-menopausal breast cancer (McPherson *et al.* 2000; Kaaks *et al.* 2002), but links between obesity and other forms of reproductive cancers are less well understood. Evidence for a link between obesity and ovarian cancer is inconsistent, but it is possible that obesity increases the risk of specific subtypes of ovarian cancer (Calle and Kaaks 2004). It may be that, as with other diseases, abdominal fat is particularly risky and there is some evidence that abdominal obesity is associated with an increased risk of breast cancer in premenopausal women (Harvie *et al.* 2003), while in men, abdominal obesity increases the risk of prostate cancer (Hsing *et al.* 2003). In contrast, in studies of premenopausal women, general obesity is associated with a reduced incidence of breast cancer, a finding that may arise at least partly because it is easier to diagnose breast cancer in lean women (Carmichael and Bates 2004).

A number of mechanisms are thought to underlie links between obesity and increased cancer risk, with insulin resistance (see p. 36) playing an important role. Thus there is increasing evidence for a link between insulin resistance, high levels of insulin and the risk of some cancers, including breast cancer and prostate cancer (Cordain *et al.* 2003; Hsing *et al.* 2003; Calle and Kaaks 2004; Renehan *et al.* 2004). One important way in which insulin resistance affects cancer risk arises because high insulin levels are associated with high levels of free insulin-like growth factor-1 (IGF-1) (Kaaks *et al.* 2002). Free IGF-1 stimulates cell division and suppresses the normal processes of cell death, increasing the risk of abnormal cell proliferation and cancer. In general, a positive energy balance, leading to obesity and hyperinsulinaemia, would also be expected to contribute to an increase in reproductive cancer risk by increasing levels of gonadal hormones (see pp. 88–93 and 96–97).

Late first birth and nulliparity as risk factors for breast cancer

The length of the interval between puberty and first pregnancy is an important risk factor for breast cancer, as is not giving birth to any children (McPherson *et al.* 2000). This is because, between puberty and first pregnancy, the epithelial cells of undifferentiated breast duct tissue are particularly susceptible to the processes that initiate the development of cancer. Full development and differentiation of breast tissue takes place only in late pregnancy. Thus late age at first birth or nulliparity increases risk because it delays the time at which breast tissue completes differentiation, with a concomitant increase in risk of a woman experiencing the initial stages of the development of breast cancer.

Exogenous oestrogens

There has been concern about the effects of synthetic or exogenous oestrogens present in the environment on the risk of developing reproductive cancers. These are chemicals manufactured by humans or found naturally in some plants that have actions that mimic the effects of endogenous oestrogens. In women exposure to synthetic or exogenous oestrogens in postmenopausal hormone replacement therapy (see Chapter 6) can increase the risk of breast cancer, although the effect appears to be generally small (Writing Group for the Women's Health Initiative Investigators 2002). Current users of combined oral contraceptives may have an increased risk of breast cancer compared with non-users, but have a reduced risk of cancer of the endometrium and ovaries (La Vecchia and Bosetti 2004), which appears to continue many years after stopping use of oral contraceptives (Hannaford *et al.* 2007). It has also been suggested that a general increase in exogenous oestrogens in the environment may have caused a recent increase in disorders of the male reproductive tract, including testicular cancer (Sharpe and Skakkebaek 1993). The focus of this concern is generally organochlorines and industrial chemicals, such as polychlorinated biphenyls (PCBs), that have been used as pesticides and have entered the human food chain. Some of these chemicals, such as dichloro-diphenyl-trichloroethane (DDT), which was used heavily in developed countries until the 1960s, are now banned there but are often still widely used in developing countries. Convincing associations have been observed between exposure to oestrogenic chemicals in pesticides and prostate cancer, but the evidence is inconclusive for other reproductive cancers (Bentley 2000; Alavanja *et al.* 2004). Exogenous oestrogens probably do not play a major role in the association between westernisation and reproductive cancers and are therefore not considered further here.

Disruption of circadian rhythm

It has been suggested that disruption of circadian rhythm may be a risk factor for cancer, particularly for breast cancer. Most interest has focused on the possible role of melatonin in affecting cancer risk. Blood melatonin concentration is normally high at night and low during the day, and its production by the pineal gland is influenced by the effect of light on the retina. As in other mammals, exposure to light during the night suppresses melatonin production, and light quality during the day also appears to affect night-time melatonin production, with some evidence that women are more sensitive to these effects than men (Stevens and Rea 2001). There is evidence to suggest that a reduction in melatonin levels at night may increase breast cancer risk, although the mechanism(s) are as yet unclear. One possibility is that melatonin may slow development and turnover of mammary cells at risk of malignant transformation (Stevens and Rea 2001). It has also been suggested that melatonin might suppress ovarian oestrogen production, although there is no clear evidence that this happens in humans (Stevens 2006). The only prospective study so far to investigate links between levels of melatonin and subsequent risk of breast cancer did not find any association, although the authors note that their limited sample size meant that they might not have been able to detect a modest association (Travis *et al.* 2004).

Explaining the high rates of reproductive cancers in western populations

Having identified the main risk factors for reproductive cancers, we must investigate the prevalence of these risk factors in affluent western societies and ask if they explain the high rates of reproductive cancers found there. Here I focus on describing and explaining levels of gonadal hormones and on fertility behaviour in affluent western societies. Levels and determinants of obesity are described in Chapter 3, but the role of obesity in affecting exposure to gonadal hormones is summarised here.

Ovarian hormone levels in western women

Ellison and colleagues have led research within anthropology on variation in levels of ovarian hormones, pioneering the assessment of ovarian steroids in saliva. This method has been particularly important for the study of gonadal hormones in women because it facilitates the collection of repeated samples from one woman. Researchers cannot generally expect large numbers of women to provide blood samples repeatedly over a menstrual

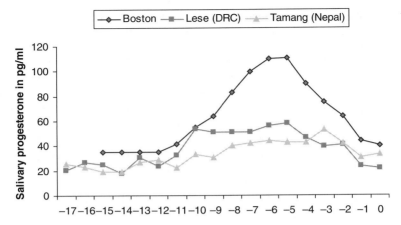

Fig. 5.3.   Interpopulation variation in luteal phase salivary progesterone levels, analysed by the same radioimmunoassay technique in the same laboratory. The data are aligned according to the start of the next menstrual cycle. Boston women ($N = 126$), Lese horticulturalists from the Democratic Republic of Congo (formerly Zaire) ($N = 47$), Tamang agropastoralists from Nepal ($N = 45$). Data used with permission from Peter Ellison and Catherine Panter-Brick. Adapted with permission from Ellison *et al.* (1993a).

cycle but it is relatively easy for women to collect saliva on a daily basis, allowing characterisation of hormone levels over the entire menstrual cycle. The method Ellison's group has made most use of is the assessment of salivary progesterone levels during the luteal phase of the menstrual cycle, when progesterone levels are at their highest (see Fig. 5.2). More recently a reliable assay for salivary oestradiol has been developed, but it is important to note that only free oestradiol levels can be assessed in saliva. This means that salivary oestradiol levels do not reflect only the level of oestradiol excreted by the ovaries, but also how much SHBG and albumin are circulating in the blood.

Ellison and colleagues showed that levels of mid-luteal phase salivary progesterone were higher in middle-class US women than in women living as subsistence agriculturalists in the Democratic Republic of Congo and Nepal (see Fig. 5.3), or in Quechua women from highland Bolivia (Ellison *et al.* 1993a). Women living in both rural and urban Bolivia also had lower mid-luteal progesterone levels than US women, with better-off urban Bolivian women having higher levels than poorer urban women (Vitzthum *et al.* 2002; Vitzthum *et al.* 2004). Consistent with these findings, white European women living in the UK had higher average salivary progesterone levels than women living in Bangladesh (Núñez-de la Mora *et al.* 2007).

Thus, there is good evidence that western women secrete higher levels of progesterone over a menstrual cycle than do women living in poorer circumstances.

Unfortunately, there is much less information available on oestrogen levels in these populations because of initial methodological difficulties in assessing salivary oestradiol. At present, the only available evidence is from a small sample of Lese subsistence horticulturalist women from the Democratic Republic of Congo, who had lower levels of salivary free oestradiol than US women (Bentley *et al.* 1998).

Important research on oestrogen levels in western women has also been conducted by epidemiologists interested in explaining the difference in breast cancer rates between women living in Asia (low rates) and women living in the west (high rates). These studies have used blood samples and examine total oestradiol levels. They suffer from the disadvantage that they generally only take one sample from each participating woman. Key *et al.* (1990) showed, using a method that involved taking a blood sample from very large numbers of women, not timed in relation to the menstrual cycle, that 35–44-year-old British women had a mean concentration of total oestradiol 36% higher than that of rural Chinese women of the same age. Other studies have also shown higher levels of total oestradiol in premenopausal women living the USA or UK than in women living in east Asia, even when the Asian women were from an affluent country such as Japan (Bernstein *et al.* 1990; Bernstein and Ross 1993).

Similar comparisons have also been made for post-menopausal women. As with pre-menopausal women, post-menopausal British women had higher serum total oestradiol levels than rural Chinese women of the same age (Key *et al.* 1990). Similarly, white US women living in a retirement community had significantly higher levels of total oestradiol and oestrone than rural Japanese women (Shimuzu *et al.* 1990).

It is clear from these findings that ovarian function in western women is not, as often considered by western biomedicine, the human 'norm'. Instead, it is likely that the high levels of ovarian hormones seen in pre-menopausal western women are, in fact, evolutionarily novel, being much higher than they are in other populations around the world and also, by extrapolation, most likely much higher than levels seen in past human populations (Ellison *et al.* 1993b). Oestrogen levels are also high, for different reasons, in post-menopausal western women. This variation in progesterone and in total and free oestradiol levels appears largely to coincide with variation in rates of reproductive cancers in women and is thought to be an important determinant of geographical patterns in reproductive cancer risk (Ellison 1999; Jasieńska and Thune 2001a).

Determinants of ovarian hormone levels

In the absence of information from hunter–gatherer populations, studies conducted amongst subsistence agriculturalist populations provide the best approximation available to us of the way in which ovarian function was regulated for most of human evolutionary history. They show that within individual and within population variation in progesterone levels is strongly influenced by energy balance (Ellison *et al.* 1993b). Energy balance refers to the relative intake and expenditure of energy. Women who are taking in more energy (eating more calories) than they expend (in metabolism and physical activity) are in positive energy balance and will be gaining weight, while women who are expending more energy than they are taking in are in negative energy balance and will be losing weight. When they are in negative energy balance, women produce relatively low levels of progesterone (Ellison *et al.* 1993b). Thus, in populations living as subsistence agriculturalists, seasonal increases in workload and weight loss have also been shown to be associated with an acute reduction in progesterone levels (Ellison *et al.* 1993b).

In women living at subsistence level any increase in energy expenditure is likely to be accompanied by weight loss. However, research with Polish farm women, who are well nourished but experience heavy workloads at certain times of the year, has shown that mid-luteal salivary progesterone levels fall in response to increased workload, despite the fact that the women show no accompanying weight loss (Jasieńska and Ellison 1998; Jasieńska and Ellison 2004). In addition, in this group a significant negative relationship was detected between levels of energy expenditure and progesterone levels.

Ellison (1990) proposed that this responsiveness of ovarian function to energy expenditure and energy balance evolved because it produced an optimal level of fecundity in relation to the energetic environment. Pregnancy, and, particularly, lactation place a huge additional energetic burden on a woman and Ellison suggests that, where energy is relatively scarce, a woman needs time between pregnancies in order to regain lost weight (Ellison 1990). Thus, it makes sense for women to secrete lower levels of ovarian hormones, reducing the likelihood of conception, when energetic conditions are difficult, that is, when a great deal of physical effort is required for subsistence or when energy put into subsistence activities is greater than that available in the diet. It is worth noting, however, that most of the evidence in favour of this hypothesis relates to variation in progesterone levels rather than to variation in levels of oestradiol.

It is plausible that the effects of energy expenditure and energy balance in adult women explain the consistent differences in ovarian function seen between populations, with western populations having particularly high levels of ovarian hormones. Women in western societies tend to be in positive energy balance, gradually gaining weight over the lifecourse on average, partly because they expend very little energy in physical activity. Both factors are expected to lead to high levels of ovarian steroids in western women if the mechanisms identified above in women living at subsistence level, or expending high levels of energy in farm-work, operate in all women (Jasieńska *et al.* 2000). Such effects are certainly seen in adult western women. Thus, for example, within western societies, athletes have lower levels of progesterone than non-athletes, even if they are not losing weight (Ellison and Lager 1986). Women who were losing moderate amounts of weight through dieting also showed reduced levels of salivary progesterone (Lager and Ellison 1990). A more recent study, focusing on determinants of free oestradiol, found that, even amongst generally well-nourished sedentary urban women, higher habitual activity is associated with lower levels of salivary free oestradiol (Jasieńska *et al.* 2006b).

In addition, Ellison has suggested that, while clearly responsive to the adult energetic environment, ovarian function may be determined partially by long-term developmental effects of the environment on ovarian function (Ellison *et al.* 1993b; Ellison 1996). Thus, girls who develop in energetically poor environments may respond by 'setting' their ovarian function at a permanently low level, reflected in late menarche and secretion of low levels of ovarian steroids for the rest of their lives, while girls developing in energetically abundant environments respond in the opposite manner. This developmental plasticity can be understood within the context provided by life history theory, which suggests that, in any environment, we should expect energy to be allocated between maintenance, growth and reproductive functions in an optimal manner (Hill and Kaplan 1999). Girls who develop in an energetically constrained environment are expected to invest energy in maintenance and growth at the expense of reproduction. Girls who develop in an energetically abundant environment, who have more than enough energy for maintenance and growth, can be expected to activate mechanisms that allow them to invest relatively heavily in reproduction. Evidence in support of this proposition comes from the finding that girls who grow fast and go through menarche relatively early (in response to an energetically abundant environment) had higher total oestradiol levels in adulthood (Apter *et al.* 1989), suggesting that they maintain a trajectory somehow set in early life. If developmental effects are indeed important, it may be that the observed differences between populations in ovarian hormone levels result from women growing up in different environments and setting their ovarian function at a particular baseline level for life, with short-term effects of energy balance and expenditure in adulthood operating to cause variation around this baseline.

A variant of this hypothesis is the suggestion that women whose own foetal development was constrained by energy availability should be more sensitive to limited energy availability as adults than women whose own foetal development was not constrained (Ellison 1996; Jasieńska *et al.* 2006a). The suggestion is that, in chronically energetically abundant environments, it is not necessary to suppress ovarian function because the prospects for child-bearing and rearing are generally good, whereas in poorer environments it is adaptive not to become pregnant when energy becomes even slightly constrained.

Migrant studies have the potential to resolve the question of whether between population differences in ovarian hormone levels result from the action of relatively short-term environmental effects on adult women or from developmental effects that determine lifetime ovarian function. In their study of Bangladeshi women, Núñez-de la Mora *et al.* (2007), found that women who had migrated as adults from Bangladesh to the UK had progesterone levels similar to those seen in women in Bangladesh. Women who had migrated before puberty showed higher progesterone levels, more similar to those of white British women. This finding is consistent with the possibility that progesterone levels may be determined to a great extent in infancy or childhood (Núñez-de la Mora *et al.* 2007). Jasieńska *et al.* (2006a) showed that Polish women who were relatively heavy for their length at birth had higher levels of salivary free oestradiol as adults. Thus there is some evidence to suggest that there are developmental influences on ovarian function in adult life, but these are not yet well understood.

The effects of energy balance and energy expenditure on ovarian function (whether during adulthood or during development) are thought to be mediated mainly by the metabolic system. Metabolic hormones, principally insulin and IGF-1, as well as the hormone leptin, signal energetic conditions to the reproductive system (Poretsky *et al.* 1999; Lipson 2001). For example, there is some evidence that increased insulin concentrations in girls may be associated with early menarche (Lipson 2001). Lipson also suggests that high levels of insulin and IGF-1 probably stimulate high levels of ovarian steroids. Several studies have shown, conversely, that levels of metabolic hormones indicating an energy deficit were associated with disrupted or reduced ovarian function (Lipson 2001).

However, there is no simple association between adiposity or insulin level in adult women and levels of steroids secreted by the ovaries. Cross-sectional studies have reported no association between insulin levels and total oestradiol sampled at the same time in pre-menopausal women (Falkner *et al.* 1999; Sutton-Tyrrell *et al.* 2005). Similarly, studies that have investigated the relationship between BMI and total oestradiol levels in pre-menopausal women have seen either no association or even, in some cases, a decrease in total oestradiol with increasing BMI (Verkasalo *et al.* 2001; Sutton-Tyrrell *et al.* 2005). Nor is there a correlation between adiposity and progesterone levels (Verkasalo *et al.* 2001; Furberg *et al.* 2005). Thus, neither current obesity nor insulin levels appear to be simply related to total oestradiol or progesterone levels. These findings do not rule out a link between metabolic hormones and ovarian function, but do suggest that the effects of the metabolic system on ovarian function are likely to be complex. Nor do they contradict Ellison's proposals about the main determinants of ovarian function, which he considers to be energy balance and energy expenditure, rather than nutritional status *per se*. The high levels of ovarian steroids seen in western populations can be explained by their generally positive energy balance and low energy expenditure, even in the absence of a simple correlation between obesity or insulin levels and ovarian function. Further, if developmental effects are key in determining the differences in levels of ovarian hormones observed between western and non-western populations, we might not expect to see simple correlations between adiposity and ovarian function in adults. It is clear that there is much yet to be discovered about mechanisms linking energetics and ovarian function.

There is, however, a strong inverse association between fatness and SHBG levels, such that obese women have low levels of SHBG, because high levels of insulin suppress production of SHBG by the liver (Poretsky *et al.* 1999; Sutton-Tyrrell *et al.* 2005). This is likely to lead to an increase in levels of free oestradiol with increasing adiposity (Verkasalo *et al.* 2001; Calle and Kaaks 2004). Such a finding was reported recently in a study of pre-menopausal Norwegian women, which found strong positive correlations of BMI, truncal fat and insulin levels with salivary free oestradiol (Furberg *et al.* 2005).

In addition, dietary intake of certain macronutrients and micronutrients has been linked to ovarian function. Thus, the differences in levels of ovarian hormones between affluent women from the east and west have been linked to dietary fat levels, which correlate positively with serum total oestradiol levels (Goldin *et al.* 1986; Wu *et al.* 1999). The fat content of the typical Japanese and Chinese diet is lower than that of typical US or UK diet (Zhou *et al.* 2003). This implies that some of the mechanisms determining ovarian function derive from the composition of the diet, not just from overall energy balance.

Another focus of attempts to explain the differences in breast cancer rates between the affluent west and east is the higher level of consumption of soy, and other foods containing phyto-oestrogens, in the east. The consumption of phyto-estrogens is now considered likely to affect oestrogen metabolism (not ovarian secretion of oestrogens), but does not appear to affect circulating levels of total oestrogens (Maskarinec *et al.* 2004).

However, phyto-oestrogens stimulate the production of SHBG and so probably protect against reproductive cancers by reducing bioavailable oestradiol levels (Adlercreutz 2002).

Amongst post-menopausal women, differences in levels of oestrogen do not reflect variation in ovarian function, since the ovaries no longer produce significant levels of hormones. As noted earlier, in post-menopausal women androgens are converted into oestrone and oestradiol in fat tissue. Thus, in post-menopausal women, levels of total oestrone and oestradiol increase with fatness (Hankinson *et al.* 1995; Verkasalo *et al.* 2001) and, as in pre-menopausal women, levels of SHBG decline with increasing fatness (Verkasalo *et al.* 2001), further contributing to higher levels of free oestradiol in fatter women. The greater fatness of western women is thought to account for some, although not all, of the differences in oestrogen levels observed in east–west comparisons of post-menopausal women (Bernstein and Ross 1993). The increase in breast cancer risk with increasing obesity in post-menopausal women is largely the result of this associated increase in oestrogen, particularly free oestradiol, levels (Endogenous Hormones and Breast Cancer Collaborative Group 2003).

In summary, given the strong association between levels of ovarian hormones and reproductive cancer risk, it is clearly very important to know why the ovaries secrete such high levels of progesterone and oestrogen during a menstrual cycle in western populations. As with so much else, the key here appears to be the western diet (specifically high overall calorie intake, a high level of fat intake and a low level of phyto-oestrogens) and patterns of physical activity. However, the mechanisms linking the energetic environment to the high ovarian hormone levels during reproductive life in western populations are mostly poorly understood. It seems possible that ovarian function may, to some extent, be determined by the energetic environment encountered early in life, with further variation caused by the energetic environment experienced during reproductive life. It is likely that the metabolic system provides an important link between the energetic environment and ovarian function, but much more work is needed to elucidate the details of these mechanisms. In post-menopausal women the situation is clearer, since greater adiposity, as seen in western women, can explain much of the higher oestrogen level observed in western women in post-reproductive life.

The assumption in these studies has been that progesterone and oestradiol levels will be regulated in similar ways, but most of the evidence on determinants of ovarian function relates to progesterone and it is possible that further study of oestradiol levels will reveal that oestradiol secretion responds differently. Since most of the hypotheses linking ovarian function with reproductive cancer risk focus on oestradiol more than on progesterone, it is important to recognise this possibility. The picture is further complicated by the need to consider determinants of SHBG levels because of the importance of SHBG in determining levels of bioavailable oestradiol.

Lifetime experience of menstrual cycles in western women

As noted above, lifetime exposure to ovarian steroids depends not only on the level of ovarian function during reproductive life, but also on the length of the reproductive lifespan, determined by age at menarche and age at menopause. In affluent western women menarche occurs earlier than in non-western populations and, as noted above, this is related to faster growth and probably to higher insulin levels in western girls. Menopause also occurs relatively late. Thus US women currently have an average age of menarche of about 12.5 years and an average age of menopause of around 51 years, while information from a number of hunter–gatherer groups suggests that average age of menarche amongst hunter–gatherers is around 16 years and average age at menopause is 47 years (Eaton *et al.* 1994; Leidy Sievert 2006, pp. 84–86).

Pregnancy and lactation are important interruptions to normal ovarian function within the reproductive lifespan, so in making the comparison between western women and hunter–gatherer women it is also important to consider patterns of fertility and breastfeeding. Women in affluent societies usually control their fertility by using contraception. In the USA at the end of the twentieth century, the total number of live births per woman averaged around 1.8 and the average duration of lactation per infant was only 3 months (Eaton *et al.* 1994). In contrast, amongst present-day hunter–gatherers women have an average of about 5 live births during their reproductive lifetimes (Bentley *et al.* 1993; Eaton *et al.* 1994) and lactational amenorrhoea is thought to last up to 4 years after each birth, with an average duration of breastfeeding per child over several societies of 2.9 years (Konner and Worthman 1980; Eaton *et al.* 1994). This mechanism is thought to have evolved to promote appropriate birth-spacing, so that the first child is well established before the next is conceived. Mothers with poor energy balance have longer periods of lactational amenorrhoea, and thus longer inter-birth intervals, partly because of their own poor nutritional status, but also because of more intense suckling by the infant (Bentley *et al.* 2001). Thus women in the USA spend much less of their reproductive lifetime either pregnant or, particularly, breastfeeding. Even when they do breastfeed for an extended time, they are less likely to experience amenorrhoea than a less well-nourished hunter–gatherer woman.

A study of the Dogon of Mali provides particularly detailed information on the experience of menstrual cycles in a group of people living at subsistence level, although they are subsistence agriculturalists and not hunter–gatherers. Like humans during most of evolutionary history, they do not use contraception although they have a higher fertility rate than is thought to be typical of hunter–gatherers, with an average of 8.6 live births experienced per woman (Strassmann 1999).

Strassmann was able to study menstruation amongst the Dogon mainly because they have strict taboos, requiring menstruating women to sleep at a menstrual hut. The purpose of this taboo appears to be to allow a woman's husband's family to know when she is cycling, pregnant or experiencing lactational amenorrhoea. This information is used to allow them to assess the paternity of children born to her (Strassmann 1999). Hormonal tests showed that attendance at the menstrual hut was indeed a reliable indicator of menstruation. Thus, Strassmann was able to collect data by visiting the menstrual hut, which she did every day for over 2 years. Using this information, she showed that menstrual cycle length in Dogon women was similar to that reported for Western women. Most importantly, she found that the median number of cycles for Dogon women over the lifespan was 109. Menstruation was a rare event for these women during the primary child-bearing years (20–34 years), because pre-menopausal women spent most of their time in lactational amenorrhoea. The figure of 109 contrasts with information from an American physician, who kept a record of her cycling throughout her reproductive years. She had three live births and 355 menstrual cycles (Treloar *et al.* 1967). Eaton *et al.* (1999a) estimated that American women experience an average number of 450 cycles during a lifetime, more than four times the number seen in Dogon women.

Thus early menarche and late menopause, together with the small numbers of pregnancies and short time spent breastfeeding by the average western woman, lead to the experience of a large number of menstrual cycles compared to the number likely to have been experienced by women during most of human evolutionary history. This contributes to high lifetime exposure to endogenous oestrogen and progesterone and thus to the high risk of reproductive cancers in western women.

Testosterone levels in western men

Levels of salivary free testosterone have been studied in a number of non-western populations, allowing comparisons with US men (Fig. 5.4). The best evidence comes from the results of a group of studies that examined free salivary testosterone levels in men in four populations, the US, the Democratic Republic of Congo, Nepal and Paraguay (Ellison *et al.* 2002). The data for the US, Congo and Nepal come from the same populations that Ellison and colleagues investigated in their work on ovarian hormone levels. The data for Paraguay come from Ache hunter–gatherers and provide the best available estimate of what levels might have been for most of human evolutionary history. Mean salivary testosterone levels for men in the four

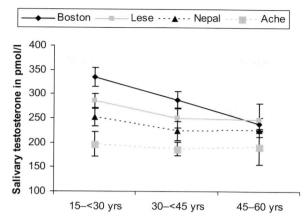

Fig. 5.4. Mean morning salivary testosterone levels by age in men from four populations, analysed by the same radioimmunoassay technique in the same laboratory. US men from Massachusetts ($N = 106$), Lese horticulturalists from the Democratic Republic of Congo ($N = 33$), Tamang agropastoralists from Nepal ($N = 39$) and Ache hunter–gatherers from Paraguay ($N = 45$). Data taken from Ellison *et al.* (2002) and used with permission from Peter Ellison, Gillian Bentley, Catherine Panter-Brick and Rick Bribiescas.

populations were significantly different in younger men, but not in older men. In younger (15–45-year-old) men levels were much higher in the US than in the other populations, with the lowest levels seen in the Ache. In older (45–60-year-old) men, there was no significant difference between the populations. There was a big decline in testosterone levels with age in the US, and a smaller decline with age in the Congo, but in Nepal and in the Ache there was no significant decline in testosterone levels with age. The pattern of variation in younger men coincides with patterns of prostate cancer incidence, which is particularly high in the US.

The question remains as to how much of the difference in free testosterone levels observed between the US and other populations derives from variation in total testosterone levels secreted by the testes and how much from variation in SHBG. Information from another small set of studies that have compared total testosterone levels in serum in western and non-western populations can help answer this question, although unfortunately these studies tend to focus on older men, and when they include younger men they do not make separate comparisons for older and younger men. Populations compared include Australians and Chinese (Jin *et al.* 1999), Dutch and Japanese (De Jong *et al.* 1991), African-Americans and Nigerians (Jackson *et al.* 1980), and men living at different degrees of urbanisation or westernisation in South Africa (Gray *et al.* 2006). As with the salivary work, these studies found higher free testosterone levels in western than in less- or non-western populations. The results suggest further that total testosterone levels are also higher and SHBG levels are lower in western populations, so that the higher free testosterone levels seen in western populations seem to arise from a combination of the secretion of higher levels of total testosterone together with low levels of binding globulin.

Within the US population, young African-American men have been reported to have significantly higher total testosterone levels compared to non-African-Americans, variation that again coincides with variation in prostate cancer risk (Platz and Giovannucci 2004). Androgen levels also appear to be higher in African-American women during pregnancy compared to matched white women (Platz and Giovannucci 2004) (see Chapter 6 for a discussion of determinants of androgen levels in women). This may be important because it has been suggested that the high prostate cancer levels seen in African-American men may be due to foetal exposure to maternal androgens (Ross *et al.* 1995).

In summary, variation in free testosterone levels in young men appears largely to coincide with variation in rates of prostate cancer, being higher in the affluent west and lower in the east and in poorer populations. This difference in free testosterone levels may derive partly from higher total testosterone levels in western populations, but there is also evidence that lower levels of SHBG seen in western populations are an important cause of higher free testosterone levels in young men in these populations. These findings support suggestions that high levels of exposure to free androgens are likely to account for the strong association of prostate cancer with the western environment.

Determinants of testosterone levels in men

As we have seen, in contrast to ovarian hormones, the most important effects of variation in testosterone levels are not on fecundity, but on the allocation of energy to different functions in the body (see p. 82). From a theoretical perspective, Ellison and Bribiescas suggest that when chronic energetic availability is good and resources allow for increased male investment in reproductive effort as opposed to survival, men should be expected to produce more testosterone in order to improve their chances of reproductive success by increasing muscle mass and, perhaps, by increasing aggression and libido (Bribiescas 2001a; Ellison 2001). However, the costs of maintaining a high testosterone level may become too high in an environment in which energy is constrained and/or there is a large burden of infectious disease, because of the costs of maintaining these characteristics and the suppressive effects of testosterone on immune function (Muehlenbein and

Bribiescas 2005). This perspective provides a rationale for higher testosterone levels in western populations, although the mechanisms that might underlie this effect remain unexplored.

As with ovarian hormones, there is no simple relationship between adiposity and testosterone levels. There appears to be only a weak relationship, or none at all, between BMI and free testosterone (Allen and Key 2000; Gapstur *et al.* 2002). This is because there is a strong inverse association between measures of overall obesity and of abdominal obesity and both total testosterone and SHBG level (Allen and Key 2000; Gapstur *et al.* 2002). It has been suggested that the inhibition of SHBG production by high levels of insulin associated with obesity drives a reduction in total testosterone levels, as the body attempts to maintain levels of free testosterone (Allen and Key 2000; Gapstur *et al.* 2002). Thus differences in BMI or other measures of adiposity between western and non-western populations are unlikely to explain the differences in free testosterone levels observed in young men.

There is some evidence that, as with oestrogen, a high dietary fat intake, regardless of total calorie intake, can increase testosterone levels (Wang *et al.* 2005). However, some have suggested that the effect of diet on testosterone levels is not likely to be strong or important (Allen and Key 2000). Thus it is possible, but by no means certain, that the high levels of fat consumption in western populations contribute to the relatively high levels of free testosterone observed in young men. It has also been suggested that differences in diet may go some way to explaining the high testosterone levels and high rates of prostate cancer seen in African-Americans (Schröder 1996).

As in women, another focus of attempts to explain the differences in prostate cancer rates between the affluent west and east is the higher level of consumption of soy, and other foods containing phyto-oestrogens, in the east. As we have seen in women, phyto-oestrogens stimulate the production of SHBG, which in men can lead to a reduction in free testosterone levels (Griffiths *et al.* 1998).

As with ovarian hormones, it seems that the energetically abundant conditions of the affluent western environment, and the high fat and low phyto-oestrogen content of the typical western diet, lead to levels of free testosterone that are likely to be evolutionarily novel in human experience. Unfortunately, the mechanisms operating to produce this effect are even less well understood than those affecting ovarian hormone levels.

Late first birth and nulliparity in affluent western societies

In western countries menarche is, as we have seen, early, and first pregnancy is often delayed until the late twenties or later, particularly in Europe. For example, average age at first birth was 29 in Italy and Spain in 1999, an increase of 2 years from the 1990 figure (Kohler *et al.* 2002). In contrast, in hunter–gatherer societies average age at first birth is consistently just below 20 years (Eaton *et al.* 1994). An increasing number of women in affluent societies are not having children at all. For example, in England and Wales, one in ten women born in 1945 reached the end of their reproductive lives without having children, whereas the figure was one in five for women born in 1960 (Berrington 2004). It is likely that the increased interval between puberty and first birth that results plays a significant role in explaining the increasingly high rates of breast cancer in western women (Eaton *et al.* 1994; Coe and Steadman 1995; Ellison 1999). The length of this interval is probably also the most important contributor to the higher risk of breast cancer in women of high socio-economic status within western societies, since it is these women who have the oldest average age at first birth (Eaton *et al.* 1994).

Changes in light exposure

Light-induced rhythms are evolutionarily ancient and exist in virtually all species (Stevens and Rea 2001). For most of human evolutionary history humans experienced dark nights and sunlit days. In contrast, people in many populations, particularly in affluent countries, now spend much of the day and night indoors, exposed to relatively dim light with a limited spectrum. Thus, if disruption of circadian rhythm and melatonin secretion is a risk factor for breast cancer, limited exposure to sunlight may play a role in increasing risk in affluent societies.

Summary

Reproductive cancers are amongst the most important cancers in men and women, and can be characterised as diseases of the affluent western lifestyle, since they are much more common in the affluent west than in Asian countries or poorer countries. Several features of western life act together to increase the risk of these cancers, especially high lifetime production of endogenous gonadal hormones, the current pattern of childbearing in women and obesity, with important interactions between these factors. An evolutionary approach shows the striking differences between the current western profile of high gonadal hormone levels, late and reduced childbearing and obesity and the profile we deduce to have been common through most of human evolutionary history.

6 Reproductive function, breastfeeding and the menopause

In this chapter I focus on three aspects of female reproductive life that have been adversely affected by affluent western life: the relatively unrecognised problem of a high rate of impaired reproductive function and infertility in women; the consequences of a lack of breastfeeding for mothers and children; and the menopause. The focus on women is not exclusive and I also discuss briefly how men's reproductive function has been affected by life in the affluent west and consider the debate about whether an analogy can be drawn between menopause in women and the less extreme age-related decline in reproductive function experienced by men.

Impaired reproductive function

My main purpose here is to examine the effects of the high levels of obesity, insulin resistance and insulin seen in western populations (see Chapter 3) on reproductive function. I also consider some other features of the western lifestyle that may affect reproductive function.

Obesity, hyperinsulinaemia and impaired reproductive function[1]

Women living in western societies generally experience higher levels of the ovarian steroid hormones progesterone and oestradiol over a lifetime than do those living in less affluent countries and in the affluent east (see Chapter 5). In fact, ovarian hormone levels in western women are considered to represent the extreme of global variation in ovarian function. As we have seen, this is likely to be a result of mechanisms that evolved to link the availability of energy with fecundity in hunter–gatherer populations. However, the evolutionary novel levels of energy abundance found in western societies are also likely to have adverse consequences for reproductive function. The evidence for this may have been partly obscured

99

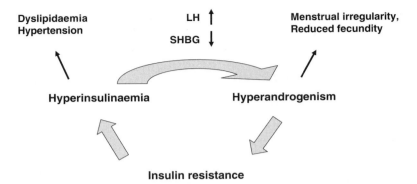

Fig. 6.1. A simplified representation of associations between insulin and androgen levels in pre-menopausal women.

by the exclusion of women who did not report regular menstrual cycles or who were obese from most of the studies of ovarian function reported in Chapter 5.

Obesity is known to be positively associated with risks of oligomenor-rhoea (infrequent menstruation), amenorrhoea (absence of menstruation) and chronic anovulation (failure to ovulate), and it is thought that hyper-insulinaemia is the main cause of these effects (Pasquali *et al.* 2003). In women, the most significant effects of abnormally high levels of insulin on levels of reproductive hormones are to increase ovarian secretion of androgens and diminish production by the liver of sex hormone-binding globulin (SHBG) (Livingston and Collison 2002). SHBG is the main binding protein for testosterone and oestradiol (see pp. 80–81 and 82) and, as we saw in Chapter 5, the level of SHBG is an important determinant of levels of free testosterone and oestradiol. Because testosterone has a relatively high affinity for SHBG, any reduction in SHBG in women increases the bioavailability of circulating testosterone in particular, and also increases the ratio of free testosterone to free oestradiol. Thus hyperinsulinaemia promotes high levels of androgens, referred to as hyperandrogenism, in women (Fig. 6.1) (Poretsky *et al.* 1999; Livingston and Collison 2002). Hyperandrogenism in turn contributes to insulin resistance, leading to a damaging cycle of interaction whereby androgen and insulin levels spiral upwards (Livingston and Collison 2002). This means that, in women who become obese and develop insulin resistance and hyperinsulinaemia, high insulin levels and high androgen levels can reinforce each other to cause significant health and reproductive problems.

Androgens have adverse effects on the growth of follicles in the ovary during the early stages of the menstrual cycle (Balen 1999), which can lead to anovulatory cycles and menstrual irregularities. Testosterone levels in girls at age 15 have been shown to be associated positively with oligomenorrhoea at age 18 (van Hooff *et al.* 2004), while higher levels of salivary free testosterone were associated with menstrual irregularities in young Canadian women (van Anders and Watson 2006). It is not surprising then that low SHBG and high free androgens are associated with reduced ability to conceive (Apter and Vihko 1990; van der Spuy and Dyer 2004).

Women severely affected by hyperandrogenism are often diagnosed with polycystic ovary syndrome (PCOS). PCOS is a syndrome of ovarian dysfunction characterised not only by high levels of androgens, but also by clinical signs of hyperandrogenism (mainly excess body hair, and often acne), menstrual irregularity and polycystic ovary morphology (enlarged ovaries with multiple cysts). In PCOS, multiple dysfunctional small cysts develop in the ovary, instead of one dominant follicle as in a normal menstrual cycle, and ovulation does not generally occur. PCOS is well known to be strongly associated with insulin resistance (Dunaif 1997). Women with PCOS may be able to conceive naturally, after treatment with drugs that reduce insulin resistance, or by *in vitro* fertilisation (IVF). However, if they do become pregnant, they have an increased risk of miscarriage, pre-eclampsia and gestational diabetes (Patel and Nestler 2006).

PCOS is the most common hormonal cause of infertility in women in the west (Balen 1999). Around 75% of women in affluent western populations who report the onset of amenorrhoea after experiencing menarche are diagnosed with PCOS (Dunaif 1997). It is reported to occur in 5%–10% of women of reproductive age in these populations (Dunaif 1997). More, perhaps 20% of women, have polycystic ovaries but not the full symptomatology of PCOS (Dunaif 1997; Hart *et al.* 2004). Thus, PCOS is an extremely prevalent disorder in the affluent west.

Populations in transition tend to have particularly high rates of insulin resistance (see Chapter 4), and we should therefore expect particularly adverse consequences for reproductive function in women in these populations. Unfortunately, information on reproductive function in populations in transition is sparse. However, one large study was conducted with the Pima, a group with high rates of obesity, insulin resistance and type 2 diabetes (see Chapter 4). The study recruited non-pregnant Pima women aged between 18 and 44 years and found a history of menstrual irregularity, defined as an interval of 3 months or more between menstrual periods after the age of 18, in 21% of women (Roumain *et al.* 1998). This compares with a rate of menstrual irregularity, defined in the same way, of 4% in a Swedish population-based study (Solomon 1999). A smaller study of British Pakistani women, another group with high levels of insulin resistance, found that androgen levels were high compared to UK women of European origin (Pollard and Unwin 2007).

Perhaps because PCOS is clearly a pathological aberration of ovarian function, it has been studied mainly by clinicians working in the field of infertility and has largely escaped the attention of human biologists with an interest in evolutionary theory. Meanwhile, despite the known links with insulin resistance, clinicians have not considered that rates of PCOS may be unusually high in western populations. This partial view results from the fact that the prevalence of PCOS has not been investigated in poorer populations, let alone in hunter–gatherers. It seems clear, however, that we should think of PCOS as one of the pathological conditions associated with the novel human experience of living in affluent western environments (Pollard and Unwin 2007). We should also consider it, like type 2 diabetes, as the tip of an iceberg. In this case the invisible part of the iceberg consists of large numbers of women with tendencies towards high androgen levels, irregular menstrual cycles and a reduced ability to conceive. These are conditions that can affect quality of life for women profoundly.

In men, obesity, hyperinsulinaemia and type 2 diabetes are strongly associated with erectile dysfunction, which is a very common cause of male infertility (Pasquali 2006). In one US study of a large cohort of men aged 40–70 years the overall prevalence of erectile dysfunction was 17%, but it increased to 45% in participants with a BMI greater than 30 (Feldman *et al.* 1994). Hyperinsulinaemia is implicated in erectile dysfunction because it damages the endothelium of blood vessels in ways that interrupt the mechanism causing the influx of blood necessary for an erection.

Thus obesity and hyperinsulinaemia are important, but sometimes unrecognised, causes of impaired reproductive function in both women and men in affluent western societies, although the mechanisms involved are very different in women and men. Unfortunately, data on changes over time in the prevalence of obesity-related fertility problems are not available, but we can surmise that they have increased and will continue to increase in prevalence worldwide.

Other features of a western lifestyle associated with impaired reproductive function

Smoking has strongly detrimental effects on fertility in women, affecting follicle development and ovulation, fertilisation and early embryo development. Furthermore, when a pregnant woman smokes, the future fertility of the foetus, as well as its general health and well-being, is put at risk, whether it is male or female (Sharpe and Franks 2002). These effects are of particular concern given the relatively slow rate of decline in smoking among younger women in the USA and elsewhere (Waldron 1991) and increases in smoking rates in women in many poorer parts of the world (Woodward *et al.* 2005).

In men, there are concerns that a sedentary lifestyle can affect fertility. The testes have to be 3–4 °C cooler than core body temperature to allow sperm production, and interference with the ability of the scrotum to cool the testes is thus another possible cause of infertility in men. Sedentary men, who, for example, spend a long time sitting in a car or at a computer, are at increased risk of a rise in scrotal temperature (Sharpe and Franks 2002).

There have been reports of a declining sperm count in American and European men in recent years, but such declines may be an artefact related, for example, to changes in assessment methods, and are anyway unlikely to have an impact on the ability of men to father children (Bribiescas 2001a). In general, sperm counts in western populations appear to be higher than in non-western populations, although, again, the differences do not seem to relate to fecundity (Bribiescas 2001a).

Thus, although impaired reproductive function is not recognised generally as a 'western' disease, it is a problem strongly associated with life in affluent western societies. This is particularly true for women, who may exacerbate the problem by delaying attempts to conceive until relatively late in reproductive life, when ovarian function is already in decline (see Chapter 5). This is not to say that our ancestors who lived as hunter–gatherers necessarily experienced a high level of fecundity, but in these populations fecundity was probably regulated in an adaptive fashion, as we saw in Chapter 5.

Breastfeeding

Breastfeeding behaviour is very different in the western world from what it would have been during most of human evolutionary history. This change has important consequences for the health of both mothers and children. Here I examine how breastfeeding practices have changed during prehistory and in historical times, and then go on to consider the health consequences of modern western infant feeding practices.

Breastfeeding in evolutionary and historical perspective

One source of information on the likely ancestral patterns of breastfeeding in the hominin evolutionary line might be to examine the determinants of infant feeding practices of other mammals, especially those of other primate species. For example, a number of proposals have been made to explain how the timing of weaning (which can refer both to the introduction of solid foods and the cessation of breastfeeding) varies across primate species. Dettwyler (1995b) found the most convincing suggestion to be that of Smith, who showed that, across primate species, age at weaning (cessation of breast-feeding) closely matches age at eruption of the first permanent molar tooth (Smith 1991). However, the uniqueness of humans, even amongst primates, makes extrapolation to humans problematic. For example, it is possible that the uniquely large brain size of anatomically modern humans makes it necessary that human infants should be weaned off breast milk earlier than other primate infants in order to satisfy the nutritional demands made by the growing brain (Kennedy 2005).

In contemporary hunter–gatherer groups breastfeeding is prolonged, with babies in almost constant contact with their mother and fed frequently, on demand (Konner and Worthman 1980). Mothers carry their babies as they forage or collect firewood (Hrdy 2000, p109). Composite data for hunter–gatherers indicate that, on average, solids are introduced into the diet at around 5 months, while breastfeeding continues until the infant is 31 months of age (Sellen and Smay 2001). !Kung babies are traditionally weaned from about $3\frac{1}{2}$ years of age (Konner and Worthman 1980). We can deduce that, for most of human evolutionary history, young children were probably breastfed until they reached around 2–3 years of age on average.

It is important to note, however, that there is considerable variation in breastfeeding behaviour across hunter–gatherer and other non-western groups. In many societies, including the !Kung, colostrum, the first milk, now known to provide important nutritional and immunological benefits to the newborn, is considered dangerous and is not given to infants (Hrdy 2000, p. 136). Thus, it is certainly not the case that all non-western societies trad-itionally practised 'natural' or health-promoting breastfeeding behaviours and we should not assume that our *Homo sapiens* ancestors all practised optimum breastfeeding behaviours.

The earliest historical information on infant care practices comes from the civilisations of Mesopotamia, Egypt and the Levant, and records from all these societies show that children were usually breastfed for 2 to 3 years (Stuart-Macadam 1995). Similar periods of breastfeeding are also mentioned in texts from other ancient civilisations (Stuart-Macadam 1995). One text from ancient India recommends breast milk as the only food for babies until the end of the first year, then milk and solid food until the second birthday, then gradual withdrawal of breast milk (Stuart-Macadam 1995).

The history of breastfeeding in the west is also informative. Historical sources from Europe show that, even when a mother did not breastfeed her own baby, the only alternative found in most societies was breastfeeding by another woman. From the eleventh century, it was common for wealthy families to employ such wet nurses, but in most of the population mothers breastfed their own children, probably for at least 1 year to 3 or more years (Fildes 1995). From the fifteenth century, babies in some areas of northern Europe were not breastfed, but given animal milks or soft preparations of cereals, but this was only possible in colder areas where these alternative foods could be stored for reasonable periods of time (Fildes 1995). Industrialisation saw increasing use of alternatives to breastfeeding in the general population. Women who moved to the new manufacturing towns were likely to supplement their breast milk with other foods and to breastfeed for only a few months, often because they needed to return to work and left their children in the care of old women or young girls (Fildes 1995).

This phenomenon is still important today in the west, since in many western countries a large proportion of women of reproductive age work outside the home. Employment often makes it difficult for women to comply with current World Health Organization recommendations that babies should be exclusively breastfed for the first 6 months of life (Kramer and Kakuma 2002). Recent studies in the USA show that initiation of breastfeeding is not affected by an intention to return to work after childbirth, but that the duration of breastfeeding is strongly influenced by maternal employment (Meek 2001). In one study, women working full-time at 3 months after childbirth breastfed for an average of 9 weeks less than women who did not work (Fein and Roe 1998). Part-time work appears to curtail breastfeeding less (Meek 2001). Provisions for maternity leave differ considerably among affluent western countries, with shorter maternity leave associated with shorter duration of breastfeeding (Meek 2001). Women in many western countries also face other barriers to breastfeeding, including perceived disapproval of breastfeeding in public places (Dettwyler 1995a). It is not difficult to understand how bottle feeding became more prevalent and why it remains popular in the affluent west. However, promotion of breastfeeding has led to increases in rates of breastfeeding in many western countries in recent years. In 2001, 70% of US mothers initiated breastfeeding, while 33% were breastfeeding at 6 months, the highest recorded figures up to that time (Ryan *et al.* 2002). In England and Wales in 2005, 77% of mothers initiated breastfeeding, also reflecting an upwards trend (Bolling 2006).

Health consequences of little or no breastfeeding for the mother

During intensive breastfeeding, ovarian function is usually suppressed and women do not experience menstrual cycles. As noted in Chapter 5, this mechanism is thought to have evolved as a result of its advantages in ensuring that a woman who is expending considerable energy on producing milk for a baby does not become pregnant again too soon. In the type of environment in which humans spent most of their evolutionary history this mechanism works to the advantage of both mother and her offspring, allowing the mother enough time between pregnancies to recover her own energy stores, and making sure that the existing baby can be properly nourished, without competition for the mother's resources from a developing foetus. This mechanism undoubtedly remains important in protecting the health of mothers and babies living in poor environments.

However, western women experience pregnancy and breastfeeding in a very different environment from that in which humans evolved, one in which energetic constraints are unlikely to impact adversely on women's health. As we have seen, the opposite is more likely to be true, with an overabundance of energy leading to weight gain. In affluent western populations women often gain weight during pregnancy and in this situation a failure to breast-feed contributes to a failure to lose weight and return to pre-pregnancy weight (Heinig and Dewey 1997). When pregnancy is associated with a sustained weight gain in the mother, there are important health implications for the mother, but such weight gain is also associated with a higher risk of adverse pregnancy outcomes. A Swedish study showed that women whose BMI increased by 3 or more units after a first pregnancy had an increased risk during subsequent pregnancy of pre-eclampsia, gestational hypertension, gestational diabetes, caesarean delivery and stillbirth compared to women who did not gain weight or whose BMI did not increase so much (Villamor and Cnattingius 2006). Method of infant feeding was not assessed so it is not possible to establish how much it might have contributed to weight change in this group. Nevertheless, it is clear that, in affluent women, a decision not to breastfeed can lead to health problems associated with weight gain.

Another important consequence of choosing not to breastfeed, curtailing breastfeeding (or not having any children to breastfeed) in affluent western populations is an increased risk of breast cancer and ovarian cancer and probably an increased risk of endometrial cancer in future years (Rosenblatt *et al.* 1995; Zografos *et al.* 2004). It has also been shown that breastfeeding protects the mother against subsequent development of insulin resistance, although the mechanisms involved here are not understood (Taylor *et al.* 2005).

Health consequences of little or no breastfeeding for the child

For the child, an increasing number of disadvantages of having been breastfed for only a short time or not having been breastfed at all have been identified. The most important consequences of no breastfeeding relate to the possibilities of infection, particularly in parts of the world where infectious diseases are a major threat to the life of infants. Breast milk contains anti-bodies that allow the infant to acquire immunity to infection passively from the mother. It also contains bioactive constituents that develop, enhance and regulate the infant's immune system (Heinig 2001). Where alternative infant foods are unlikely to be sterile, and levels of infection are high, not breast-feeding increases the risk of gastrointestinal illness and death in infancy (Cunningham 1995). Even in industrialised contexts, the loss of protection from infection afforded by breastfeeding has harmful consequences, increasing rates of gastrointestinal and respiratory illness (Heinig 2001). Duration of breastfeeding is also important, with more protection for babies who are breastfed for longer. For example, a Spanish study found that over half of infant hospital admissions would be avoided if all babies were exclusively breastfed for 4 months (Talayero *et al.* 2006).

Failure to acquire maternal antibodies and other constituents of breast milk that affect the development of the immune system during breastfeeding is also thought to have longer-term effects on health. Disorders of immune regulation that are thought to be linked to lack of breastfeeding include inflammatory bowel disease, coeliac disease and multiple sclerosis (Cunningham 1995). It has also been suggested that the risk of childhood leukaemia is higher in children who have not been breastfed. This theory derives from the suggestion that leukaemia can be a consequence of a viral infection and that passive immunity acquired through breastfeeding can protect against such infections. However, the evidence for such a link is inconclusive (Guise *et al.* 2005). The consequences of having been formula-fed for the later development of allergy are discussed in Chapter 7.

In affluent western or westernising populations, not having been breastfed increases the risk of obesity in adult life, and the shorter time a person was breastfed the greater is his or her risk of obesity in later life (Harder *et al.* 2005). One possible explanation for this effect is that neonatal nutrition influences the development of neuroendocrine circuits that regulate appetite control and body weight (Harder *et al.* 2005), although these mechanisms are still poorly elaborated and understood. Also, human milk contains bioactive substances that may influence adipocyte differentiation and proliferation (Butte 2001). Not having been breastfed may also increase the risk of type 2 diabetes in children and of developing insulin resistance in adult life, although it is not yet clear why this should be so (Ravelli *et al.* 2000; Taylor *et al.* 2005). For example, a study of the Pima (see p. 57), found that babies who were not exclusively breastfed in the first 2 months of life had a much higher risk of developing type 2 diabetes than those who were (Pettitt *et al.* 1997). Not having been breastfed is also associated with a more harmful cholesterol profile in adult life, perhaps indicating permanent effects on cholesterol metabolism (Owen *et al.* 2002). People who were formula-fed have higher blood pressure than those who have been breastfed (Martin *et al.* 2005). Again, the mechanisms involved are unclear but possibilities include a higher sodium intake in babies who are bottle-fed, or the higher levels of insulin resistance in bottle-fed babies noted above.

The complexity of the effects of infant feeding on future health are illustrated in the putative associations between infant feeding and type 1 diabetes, in which the pancreas produces insufficient insulin. Most cases of type 1 diabetes are caused by an autoimmune disorder in which the immune system destroys the pancreatic β-cells that secrete insulin. Breastfeeding, through its effects on the immune system and exposure to infection, may protect against this mechanism (Cavallo *et al.* 1996; Virtanen and Knip 2003). In addition, cow's milk may have adverse effects, for example, exposure to bovine insulin may initiate the development of β-cell auto-immunity (Virtanen and Knip 2003). A fast growth rate and obesity, asso-ciated with bottle-feeding, are also thought to be risk factors for type 1 diabetes because they are associated with hyperinsulinaemia, and pancreatic cells producing a lot of insulin are more vulnerable to mechanisms leading to type 1 diabetes (Virtanen and Knip 2003).

Many studies have reported that children who are formula-fed score lower on tests of cognitive development than those who are breastfed. Outcome measures used include school marks, intelligence tests and vocabulary tests. A meta-analysis of such studies confirmed that significantly lower levels of cognitive function at 6–23 months of age were seen in formula-fed than in breastfed children and that, amongst children who were breastfed, a shorter duration of breastfeeding was associated with poorer cognitive development (Anderson *et al.* 1999). The poorer cognitive development of formula-fed children was sustained through childhood and adolescence (Anderson *et al.* 1999). There are plausible mechanisms linking feeding method to cognitive development. Most importantly, it is thought that human breast milk may support neurological development by providing long-chain polyunsaturated fatty acids (LC-PUFAs) that are major lipid components of the brain. Serum concentrations of one of these fatty acids (docosahezenoic acid (DHA)) correlate with the results of mental and psychomotor development scales (Bjerve *et al.* 1993). Formula milk is now sometimes supplemented with LC-PUFAs such as DHA, and sold at a higher price than standard formula.

For most of human evolution and history, babies are likely to have slept with their mothers to facilitate breastfeeding (McKenna *et al.* 1999; Ball 2006). Thus McKenna *et al.* (1999) suggest that the 'evolutionarily stable sleeping arrangement' consisted of infants and mothers sleeping within arm's reach (co-sleeping) and practising night-time breastfeeding. However, in many industrialised nations, the adoption of bottle-feeding and the increasing availability of houses with separate bedrooms for parents and children were linked historically with the separation of mothers and infants during sleep (Ball and Klingaman 2007). Evidence now shows that babies who are not breastfed are most likely to sleep separately from their mothers (Blair and Ball 2004).

Evolutionary biologists suggest that, for the human infant, who develops extremely slowly and is particularly dependent on his or her caregiver for a long period compared to other primate species, co-sleeping may provide a sensory link with the mother that both aids development and protects against Sudden Infant Death Syndrome (SIDS) or 'cot death', the sudden and unexplained death of an infant (McKenna *et al.* 1999). Deaths in infants aged over one month in the western world are commonly ascribed to SIDS. If co-sleeping aids development and protects against SIDS, it is possible to make the case that infant feeding method has indirect effects on health and development as a result of its consequences for sleeping arrangements and night-time parenting behaviour. However, epidemiological studies have not, so far, identified any protective effect of bed-sharing, *per se*, against SIDS (Ball and Klingaman 2007). In contrast, there has been considerable concern that bed-sharing may increase the risk of death by 'overlaying'. Such concern is likely to be misplaced in most contexts (when mothers are not affected by alcohol or drugs and do not smoke), especially as it has been shown consistently that mothers, particularly breastfeeding mothers, characteristically sleep facing their baby and curled up around it, preventing herself or anyone else lying on the baby (Ball 2006). Nevertheless, this is a controversial topic awaiting resolution.

Unfortunately, all studies of the consequences of early feeding methods are based on the effects of bottle-feeding vs. breastfeeding in babies whose parents have chosen their method of nutrition. Researchers cannot assign feeding methods randomly to babies and study the consequences, as they would when they study the effects of new drugs. This reliance on observational studies makes it difficult to separate the effects of the feeding method from other correlates of both feeding choice and health or cognitive development, such as parental socio-economic status and educational attainment. Studies often find, for example, that parents who have had more education are more likely to choose to breastfeed their infants and also to make choices that affect their health and cognitive development in other positive ways. Thus, any observed link between having been breastfed and improved health may arise because of these two separate effects of parenting behaviour. Researchers can attempt to disentangle such effects statistically, and a recent example of such an analysis suggested that apparent effects of breastfeeding on children's intelligence were accounted for entirely by the association between maternal intelligence and breastfeeding (Der *et al.* 2006). However, statistical methods cannot overcome entirely the problems posed by observational studies and in assessing the effects of breastfeeding on future wellbeing we must take into account plausible mechanisms linking feeding method and health outcomes.

In general, it is very clear that mothers who do not breastfeed their infants for a prolonged time may suffer a number of adverse consequences, while the list of potential adverse consequences for infants is even longer. It is these concerns that have driven campaigns to reinstate breastfeeding in many affluent western countries.

Menopause

As we saw in Chapter 5, menopause is experienced by women at around the age of 50 in affluent western societies. It follows a gradual decline in ovarian function with age and is experienced by women as a gradual process. However, it is defined by biomedicine as an event, which is the last menstruation, even though this is an event which is only identifiable by the subsequent lapse of 12 months without menstruation. Human females are the only primates to show such a universal and permanent cessation of reproductive function followed by a prolonged period of post-reproductive life (Leidy Sievert 2001). This is not just a phenomenon of affluent societies today; data from hunter–gatherer societies suggest that hunter–gatherers who reach the age of 45 have a life expectancy of around 20 years (Blurton Jones *et al.* 2002).

Menopause is thought to happen when the supply of oocytes, or eggs, in the ovaries is depleted. Menstrual cycles cease when there are no longer enough ovarian follicles to produce the oestrogen levels needed to maintain the endocrine axis sustaining ovarian function. The timing of this depletion depends on the number of oocytes formed in the ovary during foetal development and on the rate of loss of those eggs through the processes of degeneration and ovulation. There are around 2 million oocytes in the ovary at birth and around 400 000 at the onset of puberty. Of these, thousands degenerate and, in western industrial societies, where women typically experience around 400 menstrual cycles during a lifetime, around 400 are lost through ovulation (Leidy 1999) (see Chapter 5). The process of degeneration in the ovary, which is responsible for most of the loss of oocytes, is not well understood.

The evolution of menopause in humans

It is likely that female hominins were originally fertile until the end of the lifespan, in the same way as free-living chimpanzees usually are today, but that, when the maximum lifespan potential of humans became longer, ovarian characteristics such as oocyte number did not change to the same extent. Why should this be? That is, what were the evolutionary pressures that led to the curtailment of reproductive life in women before the end of the maximum lifespan? A number of different answers to this question have been put forward (Leidy 1999) and the dominant two explanations are summarised here.

One answer is that menopause ensures that human mothers are not too old. Mothers need to be young enough to survive the demands of pregnancy, childbirth, breastfeeding and other forms of infant care and to live long enough to care for their children during their unusually (compared to other species) long period of dependency (Leidy 1999). This is more important for mothers than for fathers, because of the greater investment made by mothers in their children (Gaulin 1980). For example, data from the Ache hunter–gatherers show that offspring survivorship is low when a mother dies in the first 5 years of her child's life (Hill and Hurtado 1991). Also, ageing oocytes are more likely to be abnormal, leading to the conception of a foetus that has poor chances of survival and reproductive success, and menopause curtails this problem (Pollard 1994).

Another hypothesis is that women experience menopause and a long post-reproductive life because in older age they are more likely to increase their reproductive success (that is, increase the copies of their own genes in sub-sequent generations) by investing in and caring for their grandchildren than by having more children of their own (Williams 1957; Hawkes 2003). Again, this applies to women rather than men because for men reproduction does not carry the same costs as it does for women. It is likely to be uniquely true for humans because of the very long period of dependency of children. This proposition is known as the 'grandmother hypothesis'. Observations of Hadza hunter–gatherers in northern Tanzania show that post-menopausal women are effective foragers, who can improve the nutritional welfare of grandchildren (Hawkes 2003). The hypothesis is also supported by findings from a detailed study conducted in rural Ethiopia, in an agropastoralist group who did not use contraception and experienced limited food availability together with high workloads, in which grandmothers were shown to have a positive effect on the survival of grandchildren (Gibson and Mace 2005).

These hypotheses are not exclusive, and in fact it has been suggested that it is the combination of both these pressures that is likely to have driven the evolution of the human menopause (Shanley and Kirkwood 2001). Human infants' relative immaturity and their long period of dependency on others mean that they have much to gain from having both a relatively young and healthy mother and contributions from grandmothers.

The menopausal transition in the affluent west

A woman approaching menopause in the affluent west is likely to anticipate a range of menopausal 'symptoms', particularly hot flushes (UK) or flashes (USA), and night sweats. A list of menopausal 'symptoms' recognised in the

Table 6.1. *A list of 'symptoms' of the menopause identified in the UK taken from the first three websites (all UK based) identified as a result of an internet search for the terms 'menopause' and 'symptoms' on 28th August 2006*

Menopausal symptoms
Hot flushes and night sweats
Fatigue
Emotional changes such as mood swings or a change in sexual interest
Depression, mood swings, forgetfulness, irritability
Feeling tense or nervous
Excitability
Attacks of panic
Difficulty in concentrating
Loss of interest in most things
Crying spells
Feeling dizzy or faint
Sleep disturbances (insomnia)
Drier skin and hair
Thinner skin
Increased growth of facial and body hair
Aches and pains in joints
Headaches
Pressure or tightness in head or body
Parts of body feeling numb or tingling
Loss of feeling in hands or feet
Breathing difficulties
Palpitations
Generalised itching
Vaginal changes – dryness, pain during intercourse, increased risk of infections
Urinary symptoms – inability to control urination (incontinence), increased frequency of urinary infections
Loss of interest in sex

UK is given in Table 6.1. The list is extensive and suggests that experiencing menopause can have a profound influence on a woman's well-being. There is little wonder, then, that the menopause is generally perceived to be an adverse experience in the UK and in the USA.

Recently, however, there has been a reassessment of the evidence that the menopausal transition is accompanied by a large number of symptoms. One longitudinal British study collected information on health problems annually, as well as on menopausal status, for a group of 1572 women, and was able to disentangle effects strongly tied to the experience of menopause from others that were likely to reflect general effects of ageing (Hardy and Kuh 2002). They found that hot flushes and night sweats were strongly associated with menopause but that psychological symptoms were associated more strongly with current life events and difficulties, such as adult children leaving home or the death of a parent, than with menopausal status. A US report that considered all the published evidence that menopausal symptoms really are caused by hormonal changes associated with loss of ovarian function, and not just by general ageing effects, found that there is strong evidence that the menopausal transition can cause hot flushes and night sweats, as well as vaginal dryness, and moderate evidence that it can cause sleep disturbance, perhaps because of night sweats. However, evidence that it causes mood symptoms, cognitive disturbances, back pain, tiredness, stiff or painful joints, urinary incontinence or sexual dysfunction was considered limited, insufficient or inadequate (NIH State-of-the-Science Panel 2005).

The National Institutes of Health (NIH) report concludes that many women have few or no symptoms associated with the menopausal transition and are not in need of medical treatment, and that the menopause has been medicalised. That is, menopause is not seen as a normal event, but as a health problem requiring medical intervention. Social scientists have recorded the history of the medicalisation of the menopause in western society and there can be little doubt that cultural pressures other than an objective interest in women's health have driven this process, including, for example, the long history of attributing women's behaviour to reproductive organs and hormones (Lock and Kaufert 2001; Meyer 2001). As Meyer (2001) notes, the word 'hysteria' derives from the Greek for uterus and was used because of the belief that unwelcome behaviour in women could be attributed to the effects of the uterus.

Post-menopause in the affluent west

The post-menopausal state is identified in the biomedical literature as causing a change in the risk of a number of diseases. These include, most importantly, a loss of bone mass and increasing risk of osteoporosis and fracture; an increasing incidence of atherosclerosis; weight gain and the development of insulin resistance; an increase in blood pressure (and a greater blood pressure response to stress); and an increased risk of Alzheimer's disease (Owens *et al.* 1993; Kuller *et al.* 1994).

Towards the end of the twentieth century, there was a move within biomedicine to consider the post-menopausal woman as being in an 'oestrogen deficient' state. This is because many of the adverse symptoms and wider health consequences of the post-menopause can be linked to the fall in oestrogen levels at menopause. Endogenous oestrogen is thought to affect many parts of the body, including the urogenital system, skin and hair, the brain, cardiovascular system and skeleton, as well as reproductive cancer risk (Palacios 1999). Other than the increased risk of reproductive cancers (see Chapter 5), these effects are positive. For example, oestrogen has positive effects on serum lipid concentrations and has direct effects on blood vessels that reduce cardiovascular disease risk (Mendelsohn and Karas 1999). Many of these beneficial effects are likely to have arisen because of the key role of oestrogen during pregnancy, when the placenta secretes both oestrogens and progesterone, with levels of oestrogen reaching around 30 times normal towards the end of pregnancy (Guyton 1986). For example, it has been suggested that the fact that oestrogen appears to buffer the sympathetic–adrenal response to stress may be important in protecting the developing foetus from the adverse consequences of stress responses, such as increased blood pressure in the mother (Kajantie and Phillips 2006).

As a consequence of the perceived detriment to health of post-menopausal 'oestrogen deficiency', many physicians felt that medical management of the menopausal transition and post-menopausal state was important (Palacios 1999). The main biomedical strategy to manage the menopause has been hormone replacement therapy (HRT), which involves giving exogenous oestrogen, usually in combination with progestins (synthetic hormones with similar effects to those of progesterone), with the aim of returning circulating oestrogen levels to those seen before menopause. HRT was found to reduce the incidence of menopausal symptoms, including a reduction in hot flushes and night sweats (Palacios 1999). There is also evidence that HRT can reduce the risk of fractures (Writing Group for the Women's Health Initiative Investigators 2002; Warren 2004). However, as would be expected, HRT does, if only slightly, increase the risk of breast cancer (Beral *et al.* 1997; Writing Group for the Women's Health Initiative Investigators 2002).

Initially, it was considered that HRT provided protection against cardiovascular disease. These findings were based on observational studies of women who had been prescribed HRT, but there was some controversy about their interpretation because it is possible that women on HRT are a self-selecting group of women who are particularly active in pursuing a healthy lifestyle, as well as medical advice and treatment. As a consequence of this controversy, two large trials were set up in the USA, randomly assigning women to take HRT consisting of oestrogen plus progestin, or a placebo. In the first study, only women with existing coronary heart disease were enrolled and, while no overall effect on coronary heart disease risk was observed, there was a small increase in the risk of a coronary heart disease event (a heart attack) during the first year of taking HRT (Hulley *et al.* 1998). The second, much larger study, was conducted with mostly healthy women. This study was stopped earlier than originally planned, after women had been in the study for an average of 5.2 years, because of a small observed increase in breast cancer risk and because there was no evidence of a protective effect with respect to cardiovascular disease (Writing Group for the Women's

Health Initiative Investigators 2002). Another part of this study investigated the effect of giving HRT consisting only of oestrogen to women who had had a hysterectomy. This part of the trial continued for longer, but was stopped after women had been participating for an average of 7.1 years because of an increase in the risk of stroke (Hendrix *et al.* 2006). These results do not necessarily tell us anything about the effects of endogenous oestrogen on women's health, but they do suggest that the use of synthetic oestrogens after the menopause is not the cure-all it was once thought to be.

As a result of these well-publicised findings, many women, in consultation with their doctors, stopped using HRT. In 2005 the NIH State-of-the-Science panel concluded that oestrogen, either by itself or with progestins, is the most consistently effective therapy for hot flushes and night sweats. However, they suggested that only women who experience severe hot flushes and/or night sweats will be willing to, or should, take on the extra risks of breast cancer and perhaps other health problems for the sake of reducing these symptoms (NIH State-of-the-Science Panel 2005).

Evolutionary perspectives on the biology of menopause

In discussions of whether we should consider the menopause as experienced by western women as a health problem or as a normal, natural process that has been subject to unhelpful medicalisation, little attention has been paid to the fact that the biological changes associated with the menopause in western women are likely to be rather different from the changes experienced by women during most of human evolutionary history. We can identify evolutionarily novel features of western life that are likely to contribute to the experience of the menopausal transition and post-menopause as problematic. Close scrutiny of the differences between experiences of menopause in the east and west provides useful information here. For example, there are many differences between US, Canadian, Japanese and Chinese women in reporting of symptoms associated with the menopause. In particular, hot flushes and night sweats are reported much less by both Japanese and Chinese women than by US or Canadian women (Lock and Kaufert 2001; Shea 2006). Given that hot flushes are the menopausal symptom most clearly associated with a decline in oestrogen levels at menopause (see above), these patterns fit nicely with the higher pre-menopausal levels of oestrogen observed in western women as compared to Chinese and Japanese women, as described in Chapter 5. Findings from a study of US women that both oestradiol levels and hot flush reports were higher in middle-aged white women than in women of Japanese and Chinese origin support this possibility (Randolph *et al.* 2003). Thus the drop in hormone levels during the menopausal transition is relatively large in affluent western women compared to women in Japan and China (or in migrants from Japan or China), and also compared to all other human populations in the past (see Chapter 5), and it is possible that this may affect the body in ways that menopause previously did not. This argument suggests that severe hot flushes are a new, typically western, menopausal experience.

High rates of osteoporosis in post-menopausal women in western populations may also be related to the fall in oestrogen level at menopause, rather than to low absolute levels after menopause. Osteoporosis is a condition in which the structure of the bones becomes porous and bone mass is low, increasing the risk of fracture. It is a major cause of morbidity in the elderly in western countries, and is more common in women than in men (De Laet and Pols 2000). Oestrogen encourages bone formation and the retention of calcium by the body and the increase in risk of osteoporosis post-meno-pausally, together with the protection against bone fractures offered by HRT, suggest that the fall in oestrogen levels after menopause is a cause. Galloway (1997) pointed out parallels between the fall in oestrogen and progesterone levels on giving birth and at menopause. She suggests that the body interprets a significant drop in ovarian steroid levels as a signal to release calcium from the bones, as happens at birth to mobilise calcium for breast milk. At menopause, this signal will be much stronger in western women than in other populations. After lactation, pre-menopausal women experience a rise in oestradiol levels that allows them to rebuild their bones, but during meno-pause oestrogen levels remain low.

Once again, context is important here. In general, conditions that show an increased incidence after menopause, such as cardiovascular disease and osteoporosis, are strongly associated with features of affluent western society, such as obesity and lack of physical activity (Leidy 1999). For example, weight-bearing physical activity, including walking and running, stimulates bone formation in children and adolescents and protects against the loss of bone mass in adults, and lack of physical activity is known to be an important risk factor for osteoporosis (Wallace and Cumming 2000). Thus, the effect of a drop in oestrogen at menopause on bone health is likely to be much worse in a sedentary western population than it would have been for our physically active ancestors. This logic suggests that we should not regard the post-menopausal condition as being inherently unhealthy – it is the combination of being post-menopausal and living an unhealthy western lifestyle that is problematic.

In summary, it seems likely that the experience of the menopause as a medical problem, and of the post-menopausal state as a risky oestrogen-deficient state, is very specific to western society, arising from a combination of living in a manner that causes high levels of ovarian function in pre-menopausal life and that increases the risk of important non-communicable diseases, with particular cultural perceptions of women and women's reproductive biology. An evolutionary perspective allows us to stop thinking about the western menopause as normal, and consider instead that the biology of the western menopause may be new in human experience.

Ageing and gonadal hormones in men

The concept used in the western world from mediaeval times to express the idea of a transition at mid-life – the climacteric – made no distinction between men and women (Lock and Kaufert 2001). Thus the notion of a significant mid-life change was originally applied to both sexes and pre-dated the more specific concept of menopause, which is normally understood to apply to women only. More recently, there have been moves to reverse this process and to apply the concept to men, labelling it male menopause, climacteric syndrome or andropause. This is despite the fact that, in women, there is no possibility of reproduction after the menopause (other than by the use of new reproductive technologies), whereas in men there is no clearly demarcated end to reproductive function.

In the *British Medical Journal* in 2000, as part of a debate on the existence of a male menopause, the following symptoms were listed as characterising the 'male climacteric syndrome': depression, nervousness; flushes and sweats; decreased libido; erectile dysfunction; easily fatigued; poor concentration and memory (Gould *et al.* 2000). The similarity with the list of menopausal symptoms given in Table 6.1 is striking. Two of the authors (Gould and Petty) argue that, while the term 'male menopause' is inappropriate, as it suggests a sudden drop in gonadal hormones as in women, nevertheless 'male climacteric syndrome' may occur in middle-aged and elderly men because of the age-related decline in testosterone production and plasma concentration (see pp. 94–95). Gould and Petty suggest that there is a threshold plasma concentration of testosterone below which symptoms may occur. They also suggest that men who present with 'menopausal' symptoms should undergo blood tests and that, if low levels of testosterone are identified, testosterone replacement therapy should be offered. More cautiously, others have called for large clinical trials on the safety and efficacy of testosterone replacement for ageing men (Cunningham 2006).

There is evidence to support the notion that the age-related decline in testosterone levels seen in western men has deleterious consequences for health, particularly in relation to bone health and loss of muscle mass. Like oestrogen, testosterone is an important determinant of bone mass, and the fall in testosterone levels with age in men correlates with bone loss. There is also evidence that, in men identified as having low testosterone levels, androgen replacement therapy increases bone mineral density and reverses muscle atrophy (Ongphiphadhanakul *et al.* 1995; Harman *et al.* 2001).

Research on between population variation in testosterone with age, however (Ellison *et al.* 2002) (see pp. 94–95), suggests that the steep age-related decline in free testosterone levels, which is generally thought to be a normal part of the ageing process in the clinical literature, may be characteristic only of western men. If this is the case, the age-related decline in testosterone should be considered an evolutionarily novel phenomenon. As with the menopause in women, it is not surpising that a decline in free testosterone with age, which was not experienced during most of human history, may have implications for health. Ellison *et al.* point out that if, for example, set points for muscle anabolism and bone mineral density are established relative to testosterone levels in young adulthood, a steep decline in testosterone from high young-adult levels might result in age-related changes in male body composition, bone mineral density and related health risks.

Thus, as with the western female menopause, a drop in gonadal hormones with age in men may be a relatively new phenomenon and may bring health costs. As with women, context is also important. For example, a decline in testosterone levels with age is much more likely to be associated with pathological changes in bone structure in a society in which men are sedentary. Furthermore, insulin resistance, which is more prevalent in western and westernising societies than elsewhere, is strongly associated with low testosterone levels (Ding *et al.* 2006), so that insulin resistance in older men may exacerbate any age-related decline in testosterone levels. Thus, we can look with some sympathy on the case that western men experience health problems as a result of an age-related decline in testosterone levels, although, as with HRT in women, this does not necessarily imply that androgen replacement therapy is the best response. Healthier eating and activity patterns would go a considerable way to alleviating these problems.

Summary

Women living in the affluent west experience the reproductive cycle, from menarche to post-menopause, very differently from women throughout most of human evolutionary history. Reproductive function in western women may be impaired because of insulin resistance in ways that are not recognised widely as being new or specific to western societies. The consequences of limited breastfeeding in western societies for the health of the mother and child are many and profound, and the trend for an increase in breastfeeding in many affluent western countries is clearly a positive one. Finally, it is difficult to disentangle the effects of changes in biology from the cultural pressures that have led to the medicalisation of the menopause, but the evidence points to biological reasons explaining part of the negative experience of the menopausal transition in western women, as well as some of the problems associated with an age-related decline in testosterone in western men.

END NOTES

[1] Most of this section is based on Pollard, T. M. and Unwin, N. (2008) Impaired reproductive function in western and 'westernizing' populations: an evolutionary approach. In *Evolutionary Medicine and Health: New Perspectives,* ed. W. Trevathan, E. O. Smith and J. McKenna. New York: Oxford University Press. By permission of Oxford University Press, Inc.

7   *Asthma and allergic disease*

The rise of asthma and allergic diseases in most affluent countries of the world over the last 40 years has been striking, and many theories have been proposed to explain the phenomenon. These theories help us to understand what has gone wrong in our bodies when we have an allergic response to apparently harmless substances, such as cat dander, pollen or peanuts. Some of these ideas incorporate evolutionary insights, in particular, the suggestion that the human immune system is now lacking contact with organisms that would previously have had an important role in directing its development.

The scale of the problem

What is asthma?

The term asthma has usually been used to describe attacks of shortness of breath and wheezing caused by swollen and inflamed airways that are prone to constrict suddenly and violently. It is more commonly seen in children than in adults. However, there is increasing concern that the term is used to describe a range of conditions that do not necessarily share common underlying pathologies (Lancet Editorial 2006). There is no single biological marker or clinical test for asthma, and symptoms, triggers and responses to treatment are variable. The most common distinction is between allergic and non-allergic asthma, with allergic asthma being more common, particularly in children. But categorising a given case of asthma, even according to this broad dichotomy, can be difficult (Wenzel 2006). The most distinctive features of allergic asthma are a positive skin-prick test, indicating the presence of antibodies to specific allergens, and a history of allergic symptoms in response to that allergen. Nevertheless, it is common to describe asthma as one disease entity. In this chapter I will use the generic term and will distinguish between allergic and non-allergic forms wherever asthma researchers have made that distinction. Because most asthma is allergic, I link it with other allergic diseases.

Whatever its underlying pathology, asthma affects the lives of sufferers in a number of harmful ways. It can place limitations on physical activity (because it is often exacerbated by activity) and constrain social, educational and occupational activities. Children can become very distressed by the disease, and treatment, often by the use of inhalers, can be difficult for them to manage. The economic cost of the disease is considerable both in terms of direct medical costs, such as hospital admissions and medications, and indirect costs, such as time lost from work (Masoli *et al.* 2004a). In severe cases, or when poorly managed, asthma can cause death, and it is estimated that asthma accounts for about 1 in every 250 deaths worldwide (Masoli *et al.* 2004a). The number of disability-adjusted life–years (DALYs) (see pp. 34) lost due to asthma worldwide has been estimated to be currently around 15 million per year, similar to the figure for diabetes (Masoli *et al.* 2004a).

What is allergic disease?

Allergies are characterised by an excessive response of some components of the immune system to a foreign substance that would normally be harmless. The most common types, apart from asthma, are allergic rhinitis or rhino-conjunctivitis (a reaction of the lining of the nose, or nose and eyes, including hay fever and allergies to dust mites and cat dander), allergic dermatitis (eczema) and food (such as peanut) allergies. Atopy, characterised by raised levels of a particular type of antibody, immunoglobulin E (IgE), underlies these allergic diseases. Atopy can be diagnosed using skin-prick tests that insert a small amount of the allergen and record the body's response by measuring any redness and weal. The atopic or allergic response is complex, but, in essence, the allergen acts to prime T-cells and ultimately to stimulate IgE production. High levels of IgE stimulate other cells to release chemicals, including histamine, that cause inflammation and the typical allergic symptoms. For example, in allergic asthma histamine released in the bronchioles of the lungs stimulates dilation of blood vessels, contraction of smooth muscle in the bronchioles and production of mucus in the airways. These changes restrict air flow and make breathing difficult.

Geographical and temporal trends in the prevalence of asthma and allergic disease

Diseases recognisable as allergies were known before the twentieth century, but appear to have been rare. For example, a very few descriptions of hay fever can be traced back to Islamic texts of the ninth century and European texts of the sixteenth century, but even in the nineteenth century hay fever was regarded as most unusual (Emanuel 1988). Emanuel, who charted the increasing rates of hay fever through the nineteenth century, referred to it as a 'post-industrial revolution epidemic'.

Over the last 40 years, there has been a sharp increase in the global prevalence of asthma, particularly in children (Braman 2006). This has been associated with an increase in atopic sensitisation (as shown, for example, by responses to skin-prick testing) and is paralleled by similar increases in other allergic disorders such as allergic rhinitis and eczema (Masoli *et al.* 2004a). Asthma is now one of the most common chronic diseases in the world, with an estimated 300 million sufferers worldwide. This figure is expected to increase to 400 million by 2025 (Masoli *et al.* 2004a). One person in 10 has asthma in North America, while in the UK it has been estimated that 1 in every 7 children aged 2 to 15 years and 1 in every 25 adults have asthma symptoms requiring treatment (Braman 2006).

The International Study of Asthma and Allergies in Childhood (ISAAC) was set up to provide systematic international comparisons of the prevalences of these diseases in childhood. The core sample of phase one of the study consisted of children aged 13–14 years who completed a simple one-page questionnaire about symptoms of asthma, allergic rhinoconjunctivitis (usually hay fever) and atopic eczema. The final sample consisted of 463 801 children from 56 countries (ISAAC Steering Committee 1998). For asthma, the highest prevalences were from centres in the UK, New Zealand, Australia and the Republic of Ireland, while the lowest prevalences were from centres in several Eastern European countries, Indonesia, Greece, China, Taiwan, Uzbekistan, India and Ethiopia. There were also wide variations within regions, especially within Europe and Asia. The highest prevalences of allergic rhinoconjunctivitis and eczema symptoms were reported from centres scattered across the world, although the countries with the lowest prevalences were generally the same as those with low prevalences for asthma (ISAAC Steering Committee 1998).

The ISAAC data have been combined with those from another, smaller-scale international survey of asthma in adults to show that 'the rate of asthma increases as communities adopt western lifestyles and become more urbanized' (Fig. 7.1) (Masoli *et al.* 2004a). Thus, for example, within Africa, asthma prevalence is highest in South Africa, the most westernised of the African countries. In developing regions asthma prevalence has risen sharply with increased urbanisation and westernisation (Braman 2006). It is in these areas that people with asthma are least likely to have access to basic asthma medications or medical care (Masoli *et al.* 2004a).

A third phase of the ISAAC study, which repeated the phase one methodology an average of 7 years afterwards, indicated overall increases in the prevalence of asthma, allergic rhinitis and eczema (Asher *et al.* 2006). The only reassuring outcome of this phase of the study was that there was a decrease in the prevalence of asthma symptoms in older children in high prevalence areas. For example, in the UK between phase 1 in 1995 and phase 2

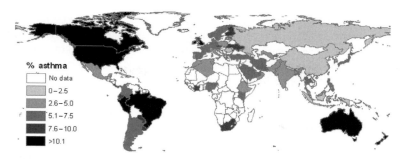

Fig. 7.1. World map showing the proportion of the population with clinical asthma by country. Adapted with permission from Masoli *et al.* (2004b).

in 2002 there were reductions in reports of wheezing or whistling in the chest in the past 12 months (a reduction of 19%) and in symptoms of allergic rhinoconjunctivitis (16%) and atopic eczema (30%) (Anderson *et al.* 2004). There were also falls in visits of older children to general practitioners for episodes of asthma and in hospital admissions (Anderson *et al.* 2004). Some of this decrease in symptoms and consultation is likely to reflect the effects of earlier detection and improved treatment of asthma in richer countries (von Hertzen and Haahtela 2005).

Allergies to peanuts and tree nuts are severe, common and long-lasting and are the leading cause of fatal allergic reactions in the USA and probably elsewhere (Sicherer *et al.* 2003). In the USA, self-reported peanut allergy doubled in children from 1997 to 2002, with a prevalence of 0.8% nationwide in 2002 (Sicherer *et al.* 2003). In the Isle of Wight in the UK, a study of 3–4 year old children in the early 1990s and again in the late 1990s showed a similar increase in self-reported peanut allergy, to 1.0% in the late 1990s (Grundy *et al.* 2002).

Exposure to allergens

The major sources of common indoor allergens are house dust mites (especially in humid areas of the world), pets (particularly cats), and cockroaches (Platts-Mills *et al.* 1997). The most important outdoor allergen is pollen. Exposure to an allergen is not enough, in itself, to cause an allergy – many exposed people remain non-allergic. However, increased exposure to indoor allergens is one of the factors contributing to the association between westernisation and allergies. For example, in Japan asthma has increased as the population has moved away from the traditional bare and well-ventilated house to western-style buildings (Cookson and Moffatt 1997). The increasing use of central heating and double-glazed windows contributes to the production of an allergenic indoor environment by creating the kind of environment that is favourable to dust mites, and reducing air flow so that allergens remain trapped in the home.

The role of air pollution

Pollution is commonly linked with asthma in the media. However, pollution is not, like asthma, a very new phenomenon (as indicated by the fact that a prohibition on burning soft coal was introduced in England in 1273 (Strassmann and Dunbar 1999)) and it is not the case that levels of asthma are always high where pollution levels are high. This is clearly seen in the global pattern of asthma prevalence observed in phase one of the ISAAC study (ISAAC Steering Committee 1998). Regions such as China and eastern Europe, with some of the highest degrees of pollution from particulates and sulphur dioxide, generally had low, albeit increasing, rates of asthma prevalence, whereas those such as western Europe and the USA with high degrees of pollution from ozone had intermediate prevalences, and some areas with the lowest degrees of air pollution, such as in New Zealand, had high prevalences of asthma.

Nevertheless, there is good evidence that pollution can contribute to an increased risk of asthma and probably hay fever. Laboratory studies have shown that particles, sulphur dioxide, nitrogen dioxide and ozone are capable of causing pathological effects on the respiratory system that could affect processes involved with the initiation or, particularly, the exacerbation of asthma and allergic rhinitis (Anderson 1997; Shusterman *et al.* 1998). For example, exposure to ozone increases the permeability of the airways to protein, which may allow allergens to penetrate into the airways (Parnia *et al.* 2002).

One area with particularly high levels of outdoor air pollution is Southern California in the USA. For this reason a study of the long-term consequences of air pollution for the respiratory health of children was undertaken there. The study found that pollutants were associated with increased new-onset asthma and asthma exacerbation and increased school absences (Künzli *et al.* 2003). In London, increases in levels of sulphur dioxide and ozone were shown to be associated with large increases in consultations with general practitioners for hay fever a few days later, particularly in children (Hajat *et al.* 2001). In China, a dramatic rise in the prevalence of asthma has been linked to severe air pollution (Watts 2006). Smoking has also been linked with incidence and severity of asthma and passive exposure to smoke increases the incidence of wheezing illness in early childhood (Anderson 1997). In general, there is agreement that air pollution can act to exacerbate existing cases of asthma and allergic rhinitis (Anderson 1997; Hajat *et al.* 2001; Parnia *et al.* 2002), but much less consensus on the role of pollution in initiating new cases of asthma.

The hygiene hypothesis

The original hygiene hypothesis: the effect of childhood infections

In a landmark study of hay fever and household size in 1989, Strachan proposed that improved hygiene in the affluent world could explain geographical and temporal trends in allergies (Strachan 1989). This hypothesis was quickly dubbed the hygiene hypothesis. In particular, the decline in childhood infections was considered important in explaining increased rates of allergic disease. Indirect evidence of an inverse relation between early childhood infections and allergies in later life comes from consistent reports of an inverse relation between number of siblings and allergy, since the presence of siblings is considered to increase exposure to infections (von Mutius *et al.* 1994; Strachan 1997). Similarly, an East German study found that, among children from small families, attending communal day-care from an early age was associated with a lower prevalence of allergy (Krämer *et al.* 1999). Furthermore, inverse associations between infections early in life and atopy amongst asthmatic children of school age have been observed (Calvani *et al.* 2002; Schaub *et al.* 2006), providing direct support for the hygiene hypothesis.

The immunological explanation underlying this association was framed in relation to T-cell subsets. T-cells are important components of the immune system and, as we have seen, play a key role in the allergic response. Two key subsets are T-helper 1 (Th1) and T-helper 2 (Th2) cells. Broadly speaking, intracellular infections by bacteria and viruses stimulate the Th1 response, whereas extra-cellular infections (including parasitic worms) stimulate the Th2 response (Yazdanbakhsh *et al.* 2002). The Th2 response is also strongly associated with allergy. When mechanisms that might explain the effects of exposure to infections in early life on allergy were first being explored, the dominant view was that bacterial and viral infections during early life direct the maturing immune system towards the Th1 and away from the Th2 pathway (Martinez 1994). A reduction in the experience of such infections in childhood was thought to result in weak Th1 imprinting and an unrestrained Th2 response, leading to an increased risk of allergy.

An evolutionary perspective provides only weak support for this original formulation of the hygiene hypothesis. As we saw in Chapter 2, most viral and bacterial infections of early childhood probably emerged only when people settled and lived together in large groups in towns and cities. Such diseases are not likely to have been a feature of human experience during the vast majority of human evolutionary history, making it difficult to argue that the immune system requires exposure to such infections in order to develop normally. Moreover, infections in early life do not consistently enhance the Th1 response and suppress the Th2 response (Schaub *et al.* 2006). Thus it seems that, while some childhood infections appear to provide protection against the development of allergic disease, links between infection and the balance between Th1 and Th2 responses are not enough to explain the recent rise in allergic disease. Therefore, we need to seek other explanations for changes in the prevalence of allergy.

The updated hygiene hypothesis: exposure to 'old friends'

More recent versions of the hygiene hypothesis have focused on the effects of an absence of exposure to non-pathogenic bacteria and helminths (nematode worms and flatworms) on the developing immune system. In evolutionary terms these versions of the hypothesis make more sense because the vast majority of previous generations of humans and their ancestors would have experienced such exposures. They are sometimes described as 'old friends' although not all, for example schistosomes, can be considered entirely friendly.

Relevant evidence comes from findings that children who grew up on livestock farms in Germany, Austria or Switzerland had lower rates of allergy than children, even in the same village, who did not live on farms (von Ehrenstein *et al.* 2000; Riedler *et al.* 2001). The suggestion is that children living on farms are exposed to something that prevents the development of an overactive Th2-dominated response to common allergens such as dust mites and pollen. The timing of exposure to the farm environment appears to be important, with the protection against allergy being strongest when exposure occurred in the first years of life (Schaub *et al.* 2006).

Interest in what might be suppressing the development of atopy in children on farms has focused on bacterial products, particularly endotoxins produced by certain types of bacteria that are commonly found in dirt and in animal faeces (Hamilton 2005). A strong inverse relationship has been found between endotoxin exposure and sensitisation to common allergens in pig farmers (Portengen *et al.* 2005) and the same association has been identified in non-farming environments (Schaub *et al.* 2006). That is, exposure to endotoxins produced by bacteria is associated with a reduction in the incidence of allergens. One type of bacteria that may be important here is saprophytic (feeding on decaying organic matter) environmental *Mycobacteria*. *Mycobacteria* are found in large quantities in mud and untreated water, but are rare on concrete or in chlorinated water (Rook *et al.* 2004). There is also a big contrast between exposure to these bacteria in developing countries, which is common, and in more affluent countries (Rook *et al.* 2004).

The suggestion is that at least some of the bacteria encountered in farm environments or in the soil would be similar to those encountered by early hominins, as a result of drinking from untreated sources of water and regular contact with soil and animals (mainly as a result of hunting) (Hamilton 2005). These bacteria are much less abundant in hygienic urban environments.

Another source of bacterial exposure for all humans is the huge reservoir of micro-organisms in the gut. The normal microbiotic inhabitants of the gut are known to act as a major stimulus of T-cell function and may play a role in the development of allergy. Studies have demonstrated differences in the composition of the gut microflora between allergic and non-allergic children, e.g. Björkstén *et al.* (1999). In particular, there is some evidence that *Lactobacillus* and *Bifidobacteria* bacteria can protect against the development of allergy (Murch 2001; Guarner *et al.* 2006).

Murch (2001) suggests that 'Probably the most striking alteration in the early immune exposures of infants since Neolithic times was the wholesale change in initial gut colonization during the past century'. The suggestion is that, given evidence that *Lactobacillus* and *Bifidobacteria* bacteria are dominant in the initial gut flora of infants in the developing world (Simhon *et al.* 1982; Murch 2001) and that infants in developing countries are colonised earlier and by a greater variety of intestinal bacteria than infants in industrialised countries (Adlerberth *et al.* 2005), we can deduce that the same is likely to have been true of most previous generations of humans. The gut microflora in infants in affluent western environments are very different, and there have also been changes in the colonisation of the gut by bacteria between the 1970s and 1990s in Europe (Adlerberth *et al.*2005).

The most important determinants of the microbiotic composition of the gut in infants include type of infant feeding, mode of delivery, infant hospitalisation and antibiotic use in infants (Penders *et al.* 2006). By the end of the first week of life, a diet of breast milk creates an environment favouring *Bifidobacteria*, but colonisation of the gut occurs differently in formula-fed infants (Fanaro *et al.* 2003). Infants born by caesarean section have a particularly delayed gut colonisation, including a slower colonisation by *Lactobacillus* and *Bifidobacteria* bacteria, than babies born by vaginal delivery (Gröland *et al.* 1999). This seems to be partly because they do not come into contact with maternal vaginal and faecal flora (Fanaro *et al.* 2003). A large proportion of babies are now born by caesarean section in many parts of the world, for example, 23% in the United Kingdom and 27% in the USA (Mayor 2005; Menacker 2005). Infants given antibiotics have a decreased number of *Bifidobacteria* (Penders *et al.* 2006). In contemporary Europe, babies born vaginally at home, and who were exclusively breastfed, appear to have what is now considered the most beneficial gut microflora (Penders *et al.* 2006). Importantly, it seems that the first bacteria to colonise the previously sterile gut of a newborn may establish a permanent niche, so that very early experience may be crucial (Murch 2001).

A number of studies have examined the relationship between helminth infection and allergy. Several studies have shown an inverse relationship between helminth infection and allergen skin test reactivity (Cooper 2004). Studies investigating links between helminth infection and actual allergic disease have produced inconsistent results in general, but do suggest a protective effect with respect to asthma (Cooper 2004). The strongest evidence for a causal association between helminth infection and atopy derives from intervention studies that have demonstrated that repeated anti-helminthic treatments administered to helminth-infected children can cause an increase in the prevalence of allergen skin test reactivity (Cooper 2004). Thus, there is good evidence that, despite the fact that helminth-infected children have often been exposed to common allergens such as house-dust mites, they are somehow protected against the full inflammatory allergic response (Yazdanbakhsh *et al.* 2002).

As we saw in Chapter 2, helminths have parasitised the hominin line from before its divergence from the great apes. In developing countries helminth infections are widespread, with first infections occurring in infancy and often persisting into adulthood. However, they are much rarer in affluent countries and there is evidence that some of this decline has taken place in recent decades; infection by pinworm (*Enterobius*) was common in European children in the middle of the twentieth century, but infestation is now less frequent and intense (Gale 2002). Thus, a reduction in helminth infection is a plausible causal correlate of the rise in the prevalence of allergy (Emanuel 1988; Gale 2002).

Given the long co-evolutionary history between humans and helminths, it is not surprising that the human immune system has mechanisms to maximise host well-being in the event of infection by helminths (Barnes *et al.* 1999). It is almost impossible for the immune system to eradicate helminth infections, but the host can try to reduce the parasite load and limit damage. It does this by producing a Th2 dominated response. However, helminth infection also appears to induce a strong anti-inflammatory regulatory network that acts to make sure that the Th2 response does not become overactive (Yazdanbakhsh *et al.* 2002). This effect may be helpful to both host and helminth, protecting both from potentially damaging chronic inflammation.

The anti-inflammatory effect induced by helminths involves another type of T-cell, only recently identified, called regulatory T-cells. As the name suggests, regulatory T-cells provide negative feedback to the immune system, suppressing inflammation via a mechanism that depends in part on production of the cytokine interleukin-10 (IL-10). As a result, both Th1 and Th2 responses are inhibited (Gale 2002; Akdis *et al.* 2005). It seems that this effect is important in preventing the overactive Th2 response associated with allergy.

Lactobacilli and *Bifidobacteria* have now also been shown to induce this anti-inflammatory network (Guarner *et al.* 2006), as do some common childhood infections. For example, infections with rhinovirus have also been associated with increases in levels of IL-10 (Schaub *et al.* 2006). It is now thought that many infectious organisms can stimulate regulatory cytokines, and that the main mechanism linking lack of infection in early life with allergy is the lack of robust anti-inflammatory regulation, with a key role for regulatory T-cells (Yazdanbakhsh *et al.* 2002; Umetsu and DeKruyff 2006). Currently, researchers are examining the role of infection and environmental exposures in stimulating a robust regulatory response. It seems likely that mechanisms giving rise to protection from allergy will differ with different infections and environmental exposures.

A link with autoimmune diseases

There are parallels between allergic disease and autoimmune disease, and the hygiene hypothesis has been extended to apply to autoimmune diseases. Like allergies, autoimmune diseases tend to have a childhood onset. As with allergies, the incidence of some autoimmune diseases, such as type 1 diabetes, systemic lupus erythematosus and Crohn's disease, has been increasing in the developed world (Dunne and Cooke 2005). Other diseases with a strong immunoregulatory component, including multiple sclerosis and inflammatory bowel disease, have also become much more common in the western world in recent years (Guarner *et al.* 2006). There is also a strong positive association between symptoms of asthma and the occurrence of type 1 diabetes at the population level (Stene and Nafstad 2001). Even more strikingly, one study showed that children with coeliac disease or type 1 diabetes were more likely to have asthma than children without these diseases (Kero *et al.* 2001).

Type 1 diabetes is probably the most important of these diseases and, as noted in Chapter 3, can arise when the immune system destroys the insulin-producing β-cells of the pancreas. Survival is only possible if the insulin can be replaced, usually by frequent injection. Although it is much less prevalent than type 2 diabetes, 5.3 million people worldwide live with the disease. A large proportion of these people are in Europe and North America, while relatively few are in Africa (Cooke *et al.* 2004). In the UK the incidence of type 1 diabetes is increasing at a rate of 2.5% per year in children aged 0–14 years (Cooke *et al.* 2004).

Unlike allergies, autoimmune disorders are characterised by a Th1 skew in the T-lymphocyte response. However, immunoregulatory effects now thought to reduce the risk of allergy, as described above, may also protect against autoimmunity. Evidence that the hygiene hypothesis is applicable to autoimmune diseases comes from studies showing that some types of infection inhibit the onset of autoimmunity. Thus infection with *Trichuris* (whipworm) is negatively correlated with the prevalence of multiple sclerosis worldwide (Fleming and Cook 2006) and seems to protect against the development of Crohn's disease, while infection with *Mycobacteria* and schistosomes appears to protect against type 1 diabetes (Dunne and Cooke 2005).

In summary, the hygiene hypothesis has stimulated valuable work into links between exposure to infectious organisms and non-harmful bacteria in the environment and allergic disease and autoimmune disease. The original formulation lacked a convincing evolutionary rationale, but the current formulation is persuasive, suggesting that lack of exposure to pathogens and environmental bacteria that would have been normal for most of human evolutionary history is associated with a failure of the immune system to develop appropriate regulatory activities. As a result, people in affluent societies are particularly vulnerable to the development of both allergic (Th2 dominated) and autoimmune (Th1 dominated) disease.

Infant feeding

A large number of adverse consequences of a lack of breastfeeding in western societies were outlined in Chapter 6, many of which derive from the fact that breastfeeding seems to play an important role in the development of the immune system. Thus we should expect infant feeding to have an impact on the likelihood of a child developing an allergy in later life. There is now good evidence that bottle-feeding increases the risk of allergic dermatitis (eczema) and some evidence that it increases the risk of asthma, especially among children with a family history of allergy (van Odijk *et al.* 2003; Friedman and Zeiger 2005). In breastfed children a shorter duration of breastfeeding is associated with a greater risk of allergy (van Odijk *et al.* 2003). This effect may be partly a result of a lack of exposure to the protective effects of breast milk as well as early exposure to potential allergens in formula milk or in solid foods. Some of the effects of infant feeding may well be related to 'hygiene' effects. For example, as we have seen, the gut flora of the breastfed and bottle-fed infant are different, with fewer *Lactobacillus* and *Bifidobacteria* in the gut of the bottle-fed child (Kemp and Kakakios 2004). Some studies have suggested that, in contrast, where the mother suffers from allergic disease, breastfeeding and the receipt of components of the allergic mother's immune system may increase the risk of allergy in the child (van Odijk *et al.* 2003).

A role for obesity

Many studies have now shown that obesity is correlated with asthma. Those who are heavier at birth or in childhood have a higher chance of developing asthma, and obese adults are more likely to have asthma than thinner adults (Flaherman and Rutherford 2006; Shore 2006). Associations between obesity and atopy more generally have not been investigated to the same extent, and the results of those studies that have assessed this relationship are inconsistent (Shore 2006).

It is possible that the association between obesity and asthma is not causal. Obesity could simply be a marker of lifestyle habits also associated with asthma. For example, we have seen that formula-fed infants are more prone to obesity in later life (Chapter 6), and they are also more prone to develop allergies. However, there are plausible mechanisms linking obesity and both non-allergic and allergic asthma.

Accumulated fat tissue can impair lung function and may increase breathlessness and thereby increase the possibility of an asthma diagnosis. Truncal obesity has mechanical effects on the lungs and on the diaphragm that can make breathing more difficult, and changed breathing patterns may then lead to a narrowing of the airways and increased contractility, as seen in asthma (Poulain *et al.* 2006; Shore 2006). Thus obesity could increase the risk of a diagnosis of asthma which is non-allergic.

As mentioned in Chapter 3, adipose tissue secretes a range of adipokines. These adipokines can cause a state of chronic inflammation, and may make the development of allergy more likely. For example, TNF-α (tumour necrosis factor-alpha) is elevated in the serum of obese humans, probably as a result of increased production in adipose tissue. TNF-α has the capacity to induce hyper-responsiveness of the airways, so increased serum TNF-α in the obese could act to promote asthma (Shore 2006). Leptin, another adipokine, is found at very high levels in obese individuals. It is known to have pro-inflammatory effects, and high serum leptin levels have been shown to be associated with asthma (Sood *et al.* 2006). In contrast, ghrelin, produced in the stomach, is found at low levels in obese individuals and counteracts many of the effects of leptin. An inverse association between levels of ghrelin and levels of IgE has been observed, which suggests that ghrelin may inhibit IgE production (Matsuda *et al.* 2006). Thus adipokines produced by adipose tissue and found at elevated levels in obese individuals may affect the immune system so as to increase the likelihood of allergy, whereas ghrelin, which is found at low levels in obese individuals, may have an immunoregulatory effect that is lost in obese individuals.

Are some groups genetically vulnerable to the development of allergy?

Most of the interest in variation in rates of allergy across groups of different ancestry has focused on variation in the prevalence of asthma. In the USA, African-American children have a high prevalence of asthma (Rodriguez *et al.* 2002) and there is also evidence that US children of Pacific Island, Filipino, Cuban or Puerto Rican ancestry have a high prevalence, while those of Mexican ancestry have a particularly low prevalence of asthma (Davis *et al.* 2006). In the UK, black people have a greater risk of developing asthma than do white people (Netuveli *et al.* 2005).

As we might expect, two types of explanation have been offered for reports of differences between ethnic groups in rates of asthma. One is that members of ethnic groups with high levels of asthma are genetically predisposed to allergy if exposed to an affluent environment; the other is that differences in the environments that these groups experience, mainly as a result of differences in socio-economic status, are enough to explain the observed differences. Given that understanding of both genetic and environmental influences on the development of allergy is still at an early stage, it is unlikely that this debate will be fully resolved in the near future. However, we can examine the evidence as it stands.

The strength of family history as a risk factor for allergy and other work on the heritability of asthma suggests that genes play an important role in determining susceptibility (Barnes 2006). Various candidate genes or gene complexes have been postulated to confer susceptibility to asthma (Hoffjan *et al.* 2003; Blumenthal *et al.* 2004; Davis *et al.* 2006). Some of these genes appear to be associated with an increased Th2 response, and others with a reduction in the activity of regulatory elements of the immune system involved in reducing inflammation (Maizels 2005; Davis *et al.* 2006). There has been much speculation that ethnic groups with different susceptibilities to asthma will have different frequencies of variants of these genes.

There is now some evidence to support this kind of speculation. One study found that African-American women were significantly more likely than White women to carry genetic variants known to increase levels of pro-inflammatory cytokines, and less likely to carry genetic variants known to increase expression of the regulatory cytokine IL-10 (Ness *et al.* 2004). Other studies have also shown that variants of genes thought to be linked to asthma are found at different frequencies in US populations of African, European, Puerto Rican and Mexican descent (Barnes 2006).

The genetic explanation has been given an evolutionary context. It has been suggested that, because *Homo sapiens* originally evolved in Africa, where helminth infections thrive, early humans and others with ancestry in tropical areas would all have had a tendency to an inflammatory Th2 dominated immune response (Barnes 2006). However, when humans spread around the world, those that came to inhabit more temperate areas, where helminth infections were far less of a threat to health, would have experienced selective pressure against a strong Th2 response, because of its liabilities in causing allergy in the absence of helminth infection (Le Souëf *et al.* 2000; Le Souëf *et al.* 2006). Similarly, the hypothesis is that the strong Th2 immune response characteristic of African and other populations originating in tropical areas is now regulated appropriately in the presence of helminth infections, but in the absence of helminth infections is inadequately regulated leading to a susceptibility to allergic disease. These ideas are supported by good evidence that gene variants that were described originally as predisposing carriers to asthma also provide resistance to helminth parasites (Maizels 2005).

The alternative suggestion is that differential environmental exposures and other non-genetic factors can explain observed differences between ethnic groups in the prevalence of asthma. Thus, for example, a national US study of adults found that the higher prevalence of asthma in Black adults appeared to be entirely explained by the greater poverty of the Black population (Rose *et al.* 2006). Poverty may increase risk of asthma by increasing exposure to pollutants, particularly tobacco smoke, and to indoor allergens such as those carried by cockroaches. Poverty is also linked with urban residence in the USA and it has been suggested that the fact that the African-American population is disproportionately concentrated into impoverished urban areas may account for much of the increased risk of asthma (Aligne *et al.* 2000). It is unlikely, however, that poverty accounts for all of the differences observed.

Most studies of differences in the prevalence of asthma or allergy by ethnic group do not take migration status into effect. Migrants from poorer countries to affluent countries have a lower risk of allergy than those born in affluent countries, consistent with the importance of early life effects on the developing immune system (Netuveli *et al.* 2005; Davis *et al.* 2006). It is possible, therefore, that differences observed between different ethnic minority groups are strongly related to differences in migration history. Different ethnic groups in countries like the USA and UK include different proportions of migrants. Differences in asthma prevalence in country of origin that may be environmentally determined are also likely to be important in determining the risk of migrant ethnic minority groups. For example, the prevalence of asthma in Puerto Rico is high, as it is in Puerto Ricans in the USA, many of whom are migrants (Pérez-Perdomo *et al.* 2003).

Statistics on the prevalence of asthma in particular ethnic groups that depend on individuals having been diagnosed with asthma, or consulting health professionals for asthma, are also affected by group differences in consulting behaviour and the differential responses of health professionals (Roberts 2002). For example, it has been hypothesised that the high asthma prevalence found among US Puerto Ricans compared with Mexican Americans is due to the better access to medical diagnosis available to Puerto Ricans (partly as a result of differentials in wealth), although not all the evidence supports this suggestion (Rose *et al.* 2006). Studies that rely on a clinical diagnosis may, therefore, have different findings from studies that use more direct forms of assessment of asthma such as self-reports of wheezing.

In summary, there is increasing evidence that genetics may play a role in causing differential vulnerability to allergy between populations usually identified as ethnic groups. The exposure of ancestral generations to helminths may well be crucial here. The greater vulnerability of some ethnic groups to asthma is probably also linked to differences in environmental exposures. Probably, the greater vulnerability of certain groups results from a combination of different genetic susceptibility and differing environmental exposures.

Summary

The strong association of asthma and other allergic diseases with an affluent lifestyle, and the recent increases in the prevalence of allergic disease, suggest that something about the affluent environment is increasing the risk of asthma and allergy. Increased exposure to indoor allergens and to pollution may play a role in increasing rates of allergy, but they cannot explain all of the observed variation in the prevalence of allergy. It seems likely that reduced exposure to some non-pathogenic bacteria and to pathogenic childhood infections, particularly helminth infections, has important effects on the developing immune system. In particular, it appears that immunoregulatory effects do not develop normally in individuals who are not exposed to such organisms. Individuals who are not breastfed may also suffer from missing input into the developing immune system from immune-related components of breast milk. Increased obesity has the potential to increase inflammation in ways that can increase vulnerability to allergic asthma and perhaps to other allergies. Finally, many genes appear to play a role in increasing vulnerability to allergy and asthma and there is evidence that some variants of these genes vary in frequency between populations. Thus, part of the reason why African-Americans, for example, have a high prevalence of asthma may be that there is a high frequency of pro-inflammatory alleles in the African-American population.

8 *Depression and stress*

Many researchers and lay members of society have expressed, or can identify with, the view that the modern western lifestyle does not nurture good mental health. The biggest foci of concern are depression and stress, both identified as increasingly important causes of ill health around the world, particularly in western societies.

According to biomedicine, depression can be an emotion, a symptom, or a disease. Lesser feelings of depression form part of the normal range of emotional experience, while clinical depression is classified as a psychiatric illness. Rates of diagnosed clinical depression have increased in the United States, Sweden, Germany, Canada and New Zealand since the Second World War and there appears to have been a decrease in the average age of onset of depression (Klerman and Weissman 1989). Projections suggest that depression is likely to be only second in importance to coronary heart disease as a cause of ill health worldwide by 2020 (Murray and Lopez 1997) and to maintain that position in 2030 (Mathers and Loncar 2006).

When people talk about being stressed, they usually mean that they are struggling to cope with the demands being made of them, and this is also the sense in which the term stress is now most often used by academics. There is a shared notion that people living in affluent industrialised societies live increasingly stressful lives, working long hours, undertaking long commutes, juggling the demands of work and family, struggling to find time for exercise, and suffering from ill health as a consequence, and that this kind of life generates more stress and stress-related disease than humans have ever felt before. Health professionals involved in primary care in the UK agree that stress is a major cause of ill health (Collins 2001).

Before applying an evolutionary perspective to these ideas, I consider definitions and understandings of the concepts of depression and stress. Evolutionary arguments have focused on explaining why depression and stress might have evolved as an adaptive response in certain circumstances, and may be elicited to a greater extent in modern western societies than in other kinds of society in ways that are no longer adaptive. Are humans adapted for a different kind of social environment and maladapted for the psychosocial environment which is part of the modern affluent lifestyle? Finally, I consider the ways in which experiences of stress and depression may elicit physiological responses and increase our risk of developing other diseases.

The influence of culture

Recent data on the international prevalence of mental health problems, including clinical depression, have been gathered using survey instruments such as the World Health Organization's Composite International Diagnostic Interview (WHO International Consortium in Psychiatric Epidemiology 2000). This kind of approach is, of course, necessary for large-scale survey work. However, as the World Health Organization acknowledges, many social scientists have raised concerns about the cross-cultural reliability of such surveys.

Any attempt to apply western definitions of depression to assess its prevalence in other cultures will not 'see' anything that falls outside that definition (Kleinman 1977). Kleinman suggests that a true comparative cross-cultural comparison must make a systematic analysis and comparison of the relevant illness categories in all the groups it considers, comparing terms for symptoms and illness labels and then translating these local cultural accounts. Nevertheless, Kleinman and many others believe that the biological component of clinical depression is important, as evidenced for example by the fact that both western and non-western patients may often be successfully treated with antidepressant drugs (Kleinman and Good 1985; Schieffelin 1985; Smith 2002, p. 140).

Kleinman illustrates his argument by considering his experiences as a psychiatrist in Taiwan. He found that many patients who displayed signs of what western biomedicine would have defined at that time as 'depressive syndrome' did not report feeling depressed and would not accept a medical diagnosis of depression. His argument is that, because mental illness is very highly stigmatised amongst the Chinese, minor psychiatric problems are usually treated as physical complaints. Furthermore, he noted that Chinese patients are unlikely to describe strong moods, whether positive or negative, because of cultural rules that prohibit such expression. Kleinman (1977) suggested that depressive syndrome is a disease (a malfunctioning biological or psychological process), which is expressed in different illness behaviour in western and Chinese cultures.

Stress is not as well established as a medically definable disease. Perhaps this has made it easier for social scientists to identify the role of social and cultural processes in defining stress as a condition. Medical anthropologists have argued that we need to understand why the notion of stress as an important cause of ill health in western societies has become so popular in social discourse and in the scientific literature (Helman 1994, p. 314). Pollock

(1988) suggests that much of the attraction of the stress concept lies in its power to reduce the arbitrariness of suffering, and that it can serve to express a variety of ideas about the social order, relating, for example, to the ways in which society might be perceived as pathogenic. Thus, one view is that the concept of stress has gained popularity in western society because of its power to express important ideas in ways that neatly intersect with the prevailing biomedical model of health.

Pollock (1988) also suggests that stress-induced illnesses are seen as the product of an 'unnatural' society, an 'unnatural' way of living. Similarly, the stress concept is often linked to the popular notion that 'modernity' is dangerous and disease-producing itself (Helman 1994). Helman cites an early example of this discourse from 1897, when the famous physician Sir William Osler described 'arterial degeneration' as resulting from 'the worry and strain of modern life' and from 'the high pressure at which men live, and the habit of working the machine to its maximum capacity'. Often, these ideas are associated with a sense of nostalgia for some more 'natural' and rural way of living.

The cultural specificity of the notion of stress can be illustrated by comparing it with a related concept, the Punjabi notion of 'sinking heart'. This concept was explored by Krause (1989), who worked with people of Punjabi origin living in Bedford in the UK. Sinking heart can be classed as an illness in which physical sensations in the heart or in the chest are experienced and are thought to be caused by excessive heat, exhaustion, worry and/or social failure. It is an example of a culturally specific (culture-bound) explanation of somatic symptoms, based on culturally specific ideas about the person, the self and the heart. Krause (1989) suggests that the western stress model and the sinking heart model are similar in that, in both, emotional and social circumstances cause, or are contributory to, physical heart distress. But, she notes, in the western model stress is primarily associated with work and the pace of modern industrial society, whereas in the Punjabi model the link is more strongly with social relationships and sinking heart experiences are attributed to problems individuals have in negotiating emotional, sexual and marital relationships and in maintaining honour and moral values. These differences appear to reflect different cultural preoccupations and they should cause us to question the easy assumption that the ideas about stress with which we are familiar in western society, particularly in English-speaking western society, should be accepted as scientific fact.

If we consider stress to be a concept that is specific to western culture, it does not make sense to ask whether people are more stressed in western societies than they are elsewhere. Alternatively, we can accept that the concept of stress, like that of depression, is heavily influenced by cultural and historical context, but also reflects some kind of biological reality (Helman 1994). This is the pragmatic approach that I adopt here.

Why do humans have the capacity to feel depressed and stressed? Evolutionary perspectives

Some evolutionary psychologists argue that selective pressures exerted during the evolutionary history of humans necessarily resulted, through the process of natural selection, in the evolution of specific cognitive mechanisms that were successful in overcoming those pressures. Thus Cosmides and Tooby (1997) write: 'natural selection slowly sculpted the human brain, favoring circuitry that was good at solving the day-to-day problems of our hunter–gatherer ancestors – problems like finding mates, hunting animals, gathering plant foods, negotiating with friends, defending ourselves against aggression, raising children, choosing a good habitat, and so on'. Others agree with the argument that our minds and their information-processing mechanisms are just as much products of the evolutionary process as are our bodies, but not with the suggestion that the brain has distinct modules shaped to carry out particular tasks (Siegert and Ward 2002; Nesse 2004). These evolutionary approaches have produced several different explanations for the capacity of humans to feel depressed.

The most common type of explanation for the capacity of humans to experience depressed mood (rather than clinical depression) focuses on the idea that depressed mood may have been adaptive in certain circumstances in our evolutionary past. For example, depressed mood may have acted as a strategy to encourage the conservation of energy and resources in hopeless situations in which it does not make sense to invest further (Nesse 2000). Another idea is that depressed mood is an unconscious, involuntary strategy adopted by individuals who find themselves in losing positions in social competition (Price and Fowkes 1997). According to this argument, depressed mood evolved as an involuntary mechanism that inhibits aggression toward more dominant individuals and signals that the depressed person is not a threat, thus reducing the likelihood of aggression or other harmful acts from the dominant individual. A slightly different formulation suggests that depressed mood minimises the risk of social exclusion by signalling low social threat and eliciting social support from others, and by inducing hypersensitivity to indicators of social threat in the self (Allen and Badcock 2006). These perspectives suggest that the capacity to feel depression evolved because it serves a useful, adaptive function.

Some have applied similar arguments to clinical depression. For example, it has been suggested that postpartum depression may have adaptive functions in helping mothers enlist investment in offspring from fathers, or in causing a mother to stop investing in an offspring who is unlikely to enhance her reproductive success, perhaps because he or she is unlikely to survive (Hagen 1999). However, support for the suggestion that such severe depression is likely to have had adaptive consequences is less strong than that for propositions that attempt to explain less severe depression.

A more specific argument has been made in relation to the capacity of humans to feel 'social anxiety', a concept which has features in common with both depression and stress. As noted in Chapter 2, hunter–gatherer societies are characteristically egalitarian and resources are generally shared amongst the group. It has been suggested that one of the mechanisms that evolved either as a cause or consequence of this feature of hunter–gatherer life is the capacity to feel social anxiety. Social anxiety refers mainly to worry about what others think, and about acceptance as part of a group (Wilkinson 1999).

The proliferation of ideas suggests that this is a complicated area, but also, perhaps, that it is relatively easy for researchers to think up good stories about the origins of depression. As Siegert and Ward (2002) have suggested, an important issue for the credibility of evolutionary psychology is how one can distinguish a sound adaptationist analysis from a well-told 'just so' story. Researchers have tried to establish rules for making such a distinction, but the area of mental health is one in which this distinction is particularly hard to establish. Specifically, Nettle (2004) suggests that adaptationist arguments for depression are flawed, partly because depression is usually associated with a poor rather than positive social outcome. For example, he suggests that depression generally impairs social functioning in ways that alienate others. However, this argument is in itself problematic, since the outcome of depression may well be different in modern society from what it would have been during most of human evolution (see below).

The most usual argument for a functional role of stress is not limited to humans and is much more clear-cut than that for depression, relating more to the physiological response to the feeling of stress than to the capacity to feel stressed. Thus, the classic 'fight or flight' response is likely to have evolved because it prepares the body for physical and mental action, and the ability to elicit this response is likely to have had clear advantages for hominins, as for other animals, for many millennia. Other components of the physiological response to feeling stressed are also likely to have been beneficial in preparing the body to overcome a threat and, subsequently, in recovering from the experience.

A mismatch between Palaeolithic minds and the contemporary western world?

It has been suggested that clinical depression is a harmful consequence of an evolved trait that was helpful in our evolutionary environment but which can have pathological consequences in the modern western environment, where either the triggers of depressed mood are experienced more frequently and/or depressed mood no longer serves an effective function, leading to a state of chronic and severe depression.

For example, the environment in which the human mind evolved was one in which an individual could expect to live with the same small group of people for most or all of his or her life. Sociality is an essential component of living as a hunter–gatherer, contributing to reproductive fitness in many ways. Thus it is probably not surprising that social isolation or loneliness is associated with clinical depression (Ernst and Cacioppo 1999). There is good evidence that people living in modern western societies have experienced increasing social isolation, caused by urbanisation, increased mobility and demographic changes (Strassmann and Dunbar 1999). Over the past two centuries, economic and social forces have encouraged people increasingly to move to take up educational and occupational opportunities, contributing to a reduction in the number of people living in close proximity to extended family (Cacioppo and Hawkley 2003). Further, the US Census Bureau has predicted that, by 2010, there will be almost 31 million people living alone in the USA. This prediction arises partly as a result of the increasing number of elderly widowed people, but is also a consequence of delayed marriage or cohabitation with a partner, delayed childbearing and high rates of divorce. For these reasons, it can be argued that people are more isolated today than ever before. Certainly, they tend to have fewer face-to-face interactions with kin, although other forms of communication, such as email, have proliferated. Such social isolation may cause depression in itself. Further, in such a context, any evolved function of depression as a social signal is not likely to be effective. If someone lives alone and has few social contacts, behaviours such as social withdrawal, helplessness and self depreciation, which may have acted as social signals in the appropriate evolutionary context, do not have any such positive effect and instead may spiral down into more depressed feelings (Allen and Badcock 2003).

It has also been suggested that the materialism and individualism characteristic of contemporary western society are inherently harmful to human psychological well-being (Eckersley 2005). Nesse (2004) points out that, in hunter–gatherer societies, most people devoted themselves to a small number of shared goals, such as gathering food and taking care of kin. In contemporary western society big rewards now go to those who allocate a large proportion of their effort to one area of life, often to education and employment. Such goals often require years of effort and may not be achieved. Nesse suggests that this can lead to individuals becoming trapped in situations that leave them depressed.

Relative poverty, which is unknown amongst hunter–gatherers, but is an increasingly important feature of modern western society, also appears to be associated with depression (Ostler *et al.* 2001). It has been suggested that this may be because the tendency to feel social anxiety, which would have been adaptive in the hunter–gatherer context, is triggered much more in unequal societies (Wilkinson 1999). Wilkinson has suggested that increased social inequality may be one of the key determinants of poor health in modern societies, and that this can largely be explained as a consequence of the 'psychological pain of low social status' (Wilkinson 1999). The daily employment experiences of those with low socio-economic status may contribute to this effect. Low status occupations are often characterised by high levels of demands but low levels of control, leading to what has been labelled job strain (Karasek 1979).

In contrast, Scheiffelin (1985) argued that the Kaluli, a small-scale society of swidden horticulturalists living in highland Papua New Guinea, are protected from clinical depression by their culture. The Kaluli have an egalitarian society, with a system of strong reciprocity, based on the exchange of support, hospitality and meat. Schieffelin also pointed out that 'notions of personal success, competence, and adequacy as measured in a competitive postindustrial social environment are not important concerns for Kaluli self-esteem' (p. 116). Further, he suggested that the Kaluli way of dealing with grief through public expression and in ceremonial contexts, drawing on social support, may be protective. His view was that, during extended fieldwork amongst the Kaluli, he only ever encountered one person whom he thought was probably suffering from severe depression.

Similarly, it is commonly asserted that stressors are more abundant in affluent western environments than they are for hunter–gatherers, or would have been for our ancestors in the Palaeolithic (Boyden 1987), with the implication that stress is likely to be a much greater problem for the inhabitants of western environments than for their ancestors or for people living in more traditional, subsistence societies today. As an example, driving is often cited as a source of stress in affluent societies, partly because of the constant vigilance required (Smith 2002, p. 33). In general, the assumption is that, in the environment in which humans spent most of their evolutionary history, life was lived at a slower pace and the experience of stress would have been relatively rare.

Living in a society experiencing the processes of westernisation and socio-economic development has also been linked with increasing mental health problems, mostly ascribed to the experience of a rapidly changing socio-cultural environment. As society changes, there can be conflict between old and new ways of thinking and behaving, and ambiguity about one's role in the social order (Janes 1990; Colla *et al.* 2006). Traditional social support networks may break down. For example, in many traditional subsistence economies, the introduction of cash crops has caused the dissolution of extended families. Thus, the introduction of onion gardens in West Africa provoked competition among relatives and reduced the number of kin who work together in collaborative units (Strassmann and Dunbar 1999).

There are certainly many aspects of the typical affluent western psychosocial environment that seem likely to contribute to depression and stress. Life in large-scale urban post-industrial societies is very different from life in the small groups in which humans lived for most of their evolutionary history and it may well be that the human mind finds this mismatch problematic. It is, however, very hard to pin down such effects. Schieffelin's work in a small-scale subsistence society backs up prevailing views about what is problematic about western society in relation to mental health. However, it is unclear how much of his findings would transfer to other small-scale subsistence societies, and how much is specific to Kaluli culture.

Mechanisms linking depression and stress with other diseases

In 1973 Harrison wrote that modern society causes problems that 'As presently recognized ... are solely psychological ills, but it is more than probable that they have organic effects ... and are adversely affecting general health'. The current consensus, which relies most heavily on results from longitudinal studies, is that, within affluent societies, there are strong links between psychosocial 'factors' (including depression and stress) and disease and mortality, especially in relation to coronary heart disease. A number of longitudinal studies have shown that those who report depression or stress are more likely than the rest of the population to go on to develop coronary heart disease (Greenwood *et al.* 1996; Hemingway and Marmot 1999). Similarly, two longitudinal studies have also shown that greater depression was associated with risk of hypertension 5–16 years later (Jonas *et al.* 1997; Davidson *et al.* 2000). There is also a reasonable consensus that depression is associated with an increased risk of diabetes and a slightly elevated risk of cancer in subsequent years (McGee *et al.* 1994; Brown *et al.* 2004). Both depression and stress have been linked to physiological changes that, if repeated, could increase vulnerability to disease, especially cardiovascular disease, but other mechanisms may also be involved.

The physiology of stress and depression

Most of our understanding of physiological mechanisms thought to be involved in linking psychosocial factors with disease comes from the tradition of studying stress in the laboratory. I will therefore consider physiological stress responses before those associated with depression.

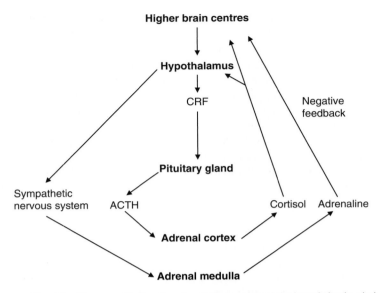

Fig. 8.1. The sympathetic–adrenal–medullary axis and the hypothalamic–pituitary–adrenal axis. ACTH = adrenocorticotropic hormone, CRF = corticotrophin releasing factor.

In laboratory studies, physiological changes in people who are experimentally stressed, for example, by undertaking difficult mental arithmetic or being asked to make a speech, are assessed. The most commonly measured responses are those of the two hormones adrenaline (known as epinephrine in the USA) and cortisol, which are controlled by separate but linked hormonal axes, the sympathetic–adrenal–medullary axis and the hypothalamic–pituitary–adrenal axis respectively (Fig. 8.1). They are often described as stress hormones. Both are secreted by the adrenal glands: adrenaline from the adrenal medulla and cortisol from the adrenal cortex. Sometimes in these studies cardiovascular responses, most commonly heart rate and blood pressure, are also assessed, but they are considered to be at least partly mediated by the stress hormone responses.

The sympathetic–adrenal–medullary axis can respond to a stressor within a few seconds (Brunner and Marmot 1999). This is the classic 'fight or flight' response, which prepares an individual to react to a threat in a number of ways, by increasing the availability of energy in the form of glucose and free fatty acids, and by increasing breathing rate, heart rate and blood pressure, all of which result in an increased supply of oxygen and sources of energy to the muscles. Adrenaline also causes platelets to become more cohesive, increasing the risk of a blood clot and thus heart attack or stroke. This effect of adrenaline may have evolved because it aided survival in dangerous situations by allowing the swift clotting of blood at wound sites (Harrison 1973). The hypothalamic–pituitary–adrenal axis responds more slowly and peak secretion of cortisol is observed after a delay of about 20–40 minutes (Pollard and Ice 2006). Stress-induced rises in cortisol support the actions of adrenaline in some ways, for example, by enhancing the production of glucose, but cortisol also acts to rein in some of the more immediate stress responses, such as increased immune activity, which might be harmful if sustained too long (Sapolsky *et al.* 2000).

The hypothalamic–pituitary–adrenal (HPA) axis is also affected in clinical depression. The dominant view has been that depressed patients tend to have high basal cortisol levels and impaired feedback in this axis (Lett *et al.* 2004). However, it is now clear that this is only true for around half of those with depression, and the picture is now generally regarded as more complicated, suggesting that HPA axis functioning in depression is best regarded as dysregulated (Lett *et al.* 2004). Although less is generally made of connections between depression and the sympathetic–adrenal–medullary axis, abnormal functioning of this axis has also been shown in depressed patients (Lett *et al.* 2004). Again, the nature of this abnormal functioning appears to be variable and is not, as yet, well understood.

Biological anthropologists have compared levels of stress hormones in western and other societies and have related the observed variation to differences in lifestyle. It is a challenge to assess levels of adrenaline and cortisol in people as they go about their everyday lives, mainly because of the difficulties of collecting bodily fluid for hormone assessment from people who are not sitting in laboratories attached to equipment to draw blood. In the case of adrenaline, research involving people in real life situations has generally relied on timed urine samples (James *et al.* 1989; Brown 2006). Urine has also been used for the assessment of cortisol, but assessment of salivary cortisol became possible in the late 1980s and is now generally preferred because it is a much simpler and more robust method (Pollard and Ice 2006). Another significant problem is that many factors other than psychosocial stress influence levels of adrenaline and cortisol, so that, even where differences are identified, it may be difficult to attribute them to psychosocial factors (Pollard 2000; Pollard and Ice 2006).

The few studies that have compared stress hormone levels between western and non-western (but unfortunately not hunter–gatherer) populations do suggest that people living in more urbanised environments excrete higher levels of urinary adrenaline than those engaged in traditional subsistence farming work. A study in Western Samoa showed that men living in the city, whether non-manual workers, students or labourers, excreted higher levels of

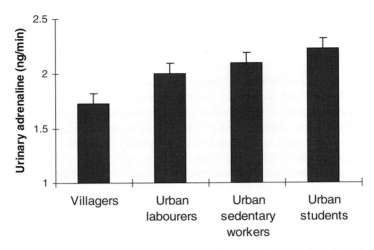

Fig. 8.2. Average urinary adrenaline excretion rates from samples collected during the morning from Western Samoan men, adjusted for urine volume, age and adiposity. Bars indicate standard errors. Sample size is approximately 30 for each group. Data from James *et al.* (1985).

urinary adrenaline than men living in villages (Fig. 8.2) (James *et al.* 1985). Similarly, Tokelauans living on Fakaofo had higher levels of urinary adrenaline than residents of Nukunonu, a nearby island less influenced by the wage economy. Still higher levels were excreted by Tokelauan migrants to New Zealand living in urban New Zealand (Jenner *et al.* 1987). In both these studies the results were explained largely with respect to working behaviour, with the suggestion that engagement in paid work, as opposed to subsistence work, was more stressful. Unfortunately, no other similar studies have been undertaken, and, given the influence of factors other than psychosocial stress on urinary adrenaline levels, these findings can only be considered suggestive. Also, the comparator group for western societies has been agriculturalists rather than hunter–gatherers (probably because there are many more subsistence agriculturalists than hunter–gatherers in the world today) and there are considerable structural differences in the ways such agriculturalists and hunter–gatherers are socially organised. Similar studies have not been conducted for cortisol.

Within western populations, urinary adrenaline levels have been shown to be higher in men at work than in the same men at leisure, but women do not consistently show such a difference (Jenner *et al.* 1980; Pollard *et al.* 1996). Workload and time pressure have been identified as key features leading to elevated adrenaline levels at work in men. For example, increased urinary adrenaline levels in Swedish bus drivers when traffic was heavier were attributed to greater perceived time pressure created by the demands of remaining on schedule (Gardell 1987). Work is also undertaken within the home and, particularly when combined with the demands of a job, is also thought to be an important stressor. This is characteristically a greater problem for women than for men because, in the west today, women are more likely than men to combine paid employment with responsibility for household duties. For example, women living in the Canadian city of Hamilton who were interviewed about the causes of stress in their lives spoke most about the difficulties of combining the many different demands they faced, usually involving family responsibilities (Walters 1993). In a USA-based study some working women gave higher ratings of stress for their time at home than at work, and these women showed less difference between their work and home blood pressures than did women more stressed by work (James *et al.* 1996).

Determinants of cortisol variation in everyday life in western societies have been more difficult to pin down, with many studies reporting null findings. However, more recent studies using improved methods in which participants have been asked to collect several samples of saliva for cortisol assessment, and to complete diary questionnaires, reporting on their experiences at the same time or around 20 minutes prior to saliva collection, have found evidence that within individuals, cortisol levels increased with stressful events and with negative mood (van Eck and Nicolson 1994; Smyth *et al.* 1998).

In summary, although there is some suggestive evidence, mostly in relation to adrenaline, we do not have an answer to the question of whether adrenaline and cortisol levels are lower in hunter–gatherers and thus, probably, were lower during most of human evolutionary history. In considering physiological stress responses in affluent western environments, researchers have focused on occupation, the conflict between the demands of work and of the home (especially for women), and, more recently, on the effects of the considerable hierarchy in socio-economic status found in western societies (see below).

How physiological changes associated with stress and depression may affect other aspects of health

The best understood example of the effects of stress on disease risk is the link between stress and cardiovascular disease (Krantz *et al.* 1988). The effects of adrenaline on blood pressure and the breakdown of lipids are thought to increase the risk of cardiovascular disease in the long term by contributing to the development of hypertension and atherosclerosis. Further, the stress response may have an immediate effect, triggering a heart attack as a consequence of changes to heart rhythm, the rupture of an atherosclerotic plaque or clotting around an atherosclerotic plaque (Kubzansky and Kawachi 2000).

Evidence of the effects of elevated levels of cortisol is provided by people with Cushing's syndrome, who secrete high levels of cortisol and also have problems with memory formation, loss of bone density and increased abdominal obesity, among other characteristics. They are also at elevated risk of diabetes, particularly type 2 diabetes (Brown *et al.* 2004). Features of Cushing's syndrome are seen in some people with clinical depression (Brown *et al.* 2004). Partly as a result of these observations, but also based on experimental studies, chronically elevated cortisol levels are now thought to increase insulin resistance, leading to an increase in insulin levels and promoting the deposition of abdominal fat (Björntorp and Rosmond 2000). For example, in one study body composition was compared in seven women with major depressive disorder and a matched control group. Although the groups did not differ in weight, BMI or total body fat, intra-abdominal fat (as measured by computed tomography scan) was more than two times greater in women with depression than in controls. Baseline cortisol levels were also significantly higher in the depressed patients and positively correlated with the amount of intra-abdominal fat (Thakore *et al.* 1997).

Repeated elicitation of adrenaline and cortisol has been described using the term 'allostatic load' (McEwen and Stellar 1993). While homeostatic responses maintain bodily systems, such as temperature, within narrow ranges, the term allostatic response has been used to describe responses, especially the physiological responses to stress outlined above, which maintain the viability of the body through change. Allostatic load is the gradual damage to the body caused by continued allostasis (McEwen and Wingfield 2003).

It has been suggested that allostatic load over a lifetime may cause physiological stress response systems to become exhausted. For example, wear and tear on the hippocampal region of the brain is thought to lead to dysregulation of the axis that controls the secretion of cortisol (Sapolsky *et al.* 1986). Such dysregulation may lead to chronically low levels of cortisol and vulnerability to problems that are normally controlled by cortisol. Chronically low levels of cortisol have been identified in patients with various illnesses, including chronic back pain and rheumatoid arthritis (Heim *et al.* 2000), but the causality of these associations is unclear. Recent research has also suggested that a disruption of the normal circadian rhythm in cortisol secretion may be associated with adverse health consequences. A study of women with breast cancer found that women who had relatively flat diurnal cycles died significantly earlier than women with more typical cycles (Sephton *et al.* 2000). The authors suggested that dysregulation of the cortisol response compromised tumour resistance by affecting immune activity.

Links between experiencing stress and the functioning of the immune system have also been explored in relation to susceptibility to infectious disease. Work by Cohen and colleagues has demonstrated that people who report more stress are also more likely to develop a clinical cold following exposure to a common cold virus (Cohen 2005). The mechanisms they have explored in most detail relate to proinflammatory cytokines, which are pro-duced in response to infection and are thought to trigger the symptoms that are associated with upper respiratory viral infections. Their results suggested that, in chronically stressed people, cortisol was less effective in regulating the proinflammatory cytokine response. In this case, then, it was not the secretion of cortisol that affected susceptibility to illness, but changes in the body's response to cortisol.

Thus, there is a variety of possible physiological mechanisms, via which depression and stress may contribute to ill health. The mechanisms are complex and still, unfortunately, less well understood than might be expected given popular acceptance of the link between stress and disease.

How stress and depression can affect health-related behaviour

In addition to associations with physiological responses that bring long-term risks of disease, the experience of stress or depression may cause changes in behaviour that have effects on health. Consistent relationships between the experience of negative emotions (anxiety, depression and anger) and increased smoking, consumption of alcohol and lack of physical activity have been reported in western societies (Kubzansky and Kawachi 2000). There is evidence that those who are depressed are less likely to be able to stop smoking (Glassman *et al.* 1990) and that depression is associated with increased alcohol use and physical inactivity (Camacho *et al.* 1991). People may also change their diet in response to stress (Steptoe *et al.* 1998). In addition, those who experience more negative emotions experience less social support, which is known to predict coronary heart disease (Hemingway and Marmot 1999). To complicate matters further, most of these behaviours have effects on the sympathetic–adrenal–medullary and hypothalamic–pitu-itary–adrenal axes, controlling the secretion of adrenaline and cortisol, respectively (Brown 2006; Pollard and Ice 2006). Thus links between depression, stress and physical ill health are complex and are not likely to be entirely mediated by physiological stress responses.

Interactions between physiological stress responses and other features of the affluent western environment

In his discussion of increased diabetes risk as a result of the misfit between human genes and the modern affluent environment, Neel (1962) makes an early mention of possible links between stress and diabetes. He notes, in particular, that 'Since the response of the adrenal cortex [the secretion of cortisol] to alarm situations is now less often followed by motor activity than in the past, one may postulate a disturbance in the physiologic balance established in the course of human evolution'. He goes on to highlight the same notion with respect to the glucose-mobilising effects of adrenaline. This is a very important point that is not given much attention in the huge literature on links between stress and ill health which developed subsequently. It seems very likely that the most significant difference in relation to physiological stress responses experienced during daily life in the affluent west today and those experienced during daily life as a hunter–gatherer in the past may not be so much one of the degree or frequency of physiological stress responses but one of context. The physiological stress response evolved to serve the function of mobilising the body (and to a certain extent the mind) when a threat was encountered. But in modern western societies stressors often do not require or evoke a physical response and most people live generally sedentary lives. The secretion of adrenaline and cortisol very often takes place against a background of a lifestyle already associated with high levels of obesity, type 2 diabetes and cardiovascular disease, as outlined in Chapter 3. It is likely that the mobilisation of free fatty acids and glucose to provide energy for the body is far more harmful in these circumstances than it would have been for most of human evolutionary history.

Psychosocial factors as potential mediators of the effects of social inequality

Within western societies it has been suggested, in line with Wilkinson's ideas on social inequality, that greater stress may help explain why people of lower socio-economic status consistently experience worse health than people of higher socio-economic status (Brunner and Marmot 1999; Steptoe and Marmot 2002). Lower levels of control available to those of lower status, in the workplace for instance, are thought to be important here. In addition, in some studies people of low occupational status have been found to encounter more frequent and negative life events and chronic stressors than people of higher socio-economic status (Matthews 2005).

For example, a study of healthy middle-aged men and women in the USA who collected data continuously for 3 days found that those with lower occupational status reported more interpersonal conflict and had a higher heart rate (although there was no such effect for blood pressure) than people of higher status (Matthews *et al.* 2000). A different US study of 24-hour urinary adrenaline levels and of salivary cortisol, assessed from samples also collected over a 3-day period, found that men and women of lower socio-economic status had higher average levels of cortisol and adrenaline (Cohen *et al.* 2006). Detailed analysis of these data suggested that the higher average levels of stress hormones were caused by people of low socio-economic status being more likely to smoke and to skip breakfast and to have less diverse social networks. Clearly, this is a complicated picture, highlighting the fact that these physiological pathways react to factors other than psychosocial influences labelled as stress.

Within westernising societies Dressler has highlighted a similar effect. He has made the valuable recommendation that we use ethnographic information to allow identification of specific, measurable features of the social environments that are important to people living in different 'modernising' environments (Dressler 1995). He has also developed a model in which he uses the methodology of cultural consensus analysis, which some cultural anthropologists have used to describe the shared beliefs of members of a particular culture, to suggest that people who do not conform to cultural expectations in ways of living show the greatest level of cardiovascular disease risk. He identifies this pathway as an important link between psychosocial experience and disease. Dressler and colleagues have shown, for example, that in urban Brazil, those whose lifestyles approximated more closely to the cultural consensus model had lower blood pressure (Dressler *et al.* 2005). Such associations bear some similarity to associations between socio-economic status, stress and health in that socio-economic status affects an individual's ability to live in the manner most valued by a particular society. Thus, in any society, individuals try to reach the defined goals, levels of prestige and standards of behaviour that the cultural group expects of its members, and failure to reach these goals may result in feelings of frustration, anxiety or depression (Helman 1994).

The suggestion that psychosocial factors may be important mediators of associations between ethnicity and health has already been discussed in Chapter 4 in relation to experiences of racism and poverty. Since ethnicity is usually strongly associated with socio-economic status, some differences in mental health across ethnic groups are likely to be determined by socio-economic status, while others may be more specifically related to ethnicity. For example, levels of depression are high among Puerto Ricans in New York, and low socio-economic status, as indicated by low education and low household income, appears to contribute to this effect (Potter *et al.* 1995). Other studies suggest a positive relationship between perceived discrimination and a diagnosis of depression (Williams *et al.* 2003). These effects would be expected to contribute to the elevated risk of coronary heart disease seen in many ethnic minority groups.

Summary

In summary, it is very difficult to test assumptions that life in western countries today is inherently more depressing or more stressful than life in non-western societies, or that the physiological effects associated with depression or stress are elicited more often than they would have been in our evolutionary past. However, there is some evidence that such contrasts are likely to exist. Within modern industrial societies there is good evidence that the experience of depression or stress can lead to physical ill-health in the long term, with the strongest link so far established being with cardiovascular disease. The critical difference between the modern western environment and the environment in which humans spent most of their evolutionary history may be that our bodies 'expect' to experience the stress response in the circumstances in which we evolved, not in the context of modern affluent western society, with its high levels of obesity and insulin resistance. It may be that this is what makes stress a risk factor for cardiovascular disease in the west.

9 Conclusion

Living in an affluent western society brings many advantages for health, notably security from hunger and from the serious infectious diseases of infancy and childhood that plagued western countries prior to the twentieth century and continue to inflict a heavy burden on populations in poorer countries today. However, as we have seen, westerners suffer from a characteristic set of relatively new non-communicable diseases, and these diseases are seen in non-western populations at increasingly high rates. In this concluding chapter I first summarise what an evolutionary perspective offers to the study of human vulnerability to western diseases. Next, I consider prospects for the future, focusing on what 'westernisation' means for the health of the millions of people subject to its influence, and on the insights that an evolutionary perspective provides in relation to possible preventive strategies.

Human vulnerability to western diseases

This book has shown that humans are vulnerable to western diseases because, as a species, we evolved in very different environments from those experienced today. We can summarise these effects in relation to obesity, in many respects the core pathology underlying western diseases. When the genus *Homo* emerged, selective pressures related especially to having a large brain led to these early hominins becoming proportionately fatter than other species of the tropical savannah. Humans also evolved under selective pressures imposed by the necessity of eating wild animals and plants and being physically active. The human genotype was 'thrifty', making effective use of scarce resources. Throughout the vast majority of human evolutionary history, those humans who effectively sought high fat, high sugar foods had the greatest reproductive success, resulting in an evolved species-wide preference for these types of foods. In contrast, those humans who now live in western or westernising environments often experience a positive energy balance as dietary supplies exceed energy requirements, making us vulnerable to obesity and to the many diseases, including type 2 diabetes, cardiovascular disease and reproductive and other cancers, associated with a

153

positive energy balance. This story encapsulates the mismatch between human biology and the affluent western environment.

However, an evolutionary perspective has more to offer than this one central insight. An emerging theme through this volume has been the importance of understanding human biology and the effects of westernisation in the context of the whole lifespan, especially in relation to developmental plasticity. It is clear that experiences during early life can have very important implications for health in later life and that an evolutionary understanding can throw important light on these effects. This is true in relation to type 2 diabetes and related disorders; being born into a relatively poor environment and later experiencing a more western environment results in a particularly high risk of disease, probably at least partly because the body has adapted to the environment experienced in early life (Chapter 4). Similarly, it seems that, during our evolutionary history, most humans experienced parasitism by worms (helminths) from early life onwards, resulting in the development of an immune system with a strong anti-inflammatory network. In the absence of such helminth infection, as in contemporary affluent societies, it seems that the lack of this regulatory control of inflammation can lead to allergy. The type of feeding experienced in infancy also has important effects for later life, so that the short duration of breastfeeding seen in affluent western societies today probably contributes to rising rates of many conditions, including allergies and autoimmune diseases. A final example relates to adverse experiences associated with the menopausal transition and post-menopausal state in women in western societies, and with an age-related decline in testosterone levels in western men. As summarised in Chapter 6, it seems likely that, for nearly all previous generations of humans, levels of these hormones were much lower in early adult life than they are in western societies today, so that we now experience a relatively steep decline with ageing, which probably has adverse consequences for our health.

Another important point to emerge repeatedly is that the obesogenic context is likely to exacerbate the effects of normal physiological processes on human health in the western context. This applies to the physiology of stress. Elevations in stress hormone levels are likely to have far worse consequences for health where, for example, levels of blood pressure and serum cholesterol are already high. Similarly, ageing effects associated with declining gonadal hormone levels, including increased vulnerability to osteoporosis, are only likely to be seriously problematic because of the fact that westerners, and increasingly others, undertake so little of the weight-bearing physical activity that is important for healthy bone formation earlier in life.

Below the species level, the controversy about whether some groups, usually identified as racial or ethnic groups, are particularly susceptible to some western diseases is a recurring theme in this volume. We saw that the thrifty genotype hypothesis does not hold up well under close examination, although the suggestion that people of European descent and perhaps some others may be particularly 'unthrifty' seems worthy of further examination. In contrast, the evidence that some groups may be more or less vulnerable than others to allergy because of greater or lesser ancestral exposure to helminths (Chapter 7) is relatively good. Given the importance of inflammation in obesity and obesity-related diseases, it is also possible that evolved differences in immune responses have implications for diseases beyond allergy.

As we have seen, and many have emphasised, the use of race as a category when examining this kind of evidence is unhelpful, though difficult to avoid (Royal and Dunston 2004). It is clear that race is not a real biological category; races are not genetically homogeneous nor distinct for most genetic variation (Tishkoff and Kidd 2004). Nevertheless, there is meaningful genetic variation between populations and this variation needs to be taken into account when considering whether some populations are particularly vulnerable to some western diseases. The challenge is to jettison traditional racial categorisations and to examine genetic variation according to likely ancestral experiences of different groups, based partly on what the study of bones, genetics and languages tells us about the history and geography of human groups (Weiss and Fullerton 2005). It is also important to understand that much genetic variation is continuous rather than discrete, and that populations or individuals differ roughly in proportion to the geographical distance between their indigenous locations, mainly because of the effects of isolation by distance (Weiss and Fullerton 2005). Selection effects can also create such clines. A good example of genetic variation that may well result from pressures imposed by continuous geographical variation, in this case in climate, is provided by genes that appear to have protected against dehydration in hotter areas for thousands of years, but now may make the descendants of these heat-adapted populations vulnerable to hypertension (pp. 63–64) (Young *et al.* 2005). Thus, we should not make the mistake of throwing the idea of genetically based susceptibility to disease, including western diseases, out with the bathwater of race.

In summary, it is quite easy to see in theory why humans as a species are vulnerable to the development of diseases in the context of a western environment. A detailed consideration of mechanisms, as presented in the body of this book, throws important light on exactly why and how this happens. Where does this leave us with respect to the future?

Western diseases: projected trends

Recent projections of global mortality suggest the proportion of deaths due to non-communicable disease will rise from 59% in 2002 to 69% in 2030 (Mathers and Loncar 2006). This is despite an increasing number of deaths caused by HIV/AIDS. The three leading contributors to illness (the burden of disease) in 2002 were perinatal conditions, lower respiratory infections and HIV/AIDS, but by 2030 they are projected to be HIV/AIDS, unipolar depressive disorders and coronary heart disease, with cerebrovascular disease (stroke) ranked sixth (Mathers and Loncar 2006). These predicted changes in disease experience mainly reflect changes in developing countries that can be crudely characterised as the epidemiological transition (see pp. 18–21), and also the ageing of the global population as fertility rates fall, so that relatively more people suffer from diseases associated with old age. It is helpful, however, to consider the future rates of western diseases separately for affluent and poorer (developing) areas of the world.

Affluent countries

Within high income countries, projections show that an ageing population is expected to lead to a gradual increase in crude death rates from many of the diseases considered in this volume, including diabetes mellitus, coronary heart disease, stroke, Alzheimer's and other dementias and reproductive cancers (Fig. 9.1). Similarly, the burden of disease as measured by Disability Adjusted Life–Years (DALYs, see p. 34) attributable to diabetes mellitus and to depression are expected to rise. Because of the strength of population ageing effects, this is true even where the age-specific rates of these diseases are declining and are expected to continue to decline, as in the case of coronary heart disease. That is, the fact that there are more older people in the population means that there will be more people who suffer and die from these diseases in absolute terms, even when a smaller proportion than previously of people of any given age are getting a particular disease.

However, except in the case of diabetes, these projections do not take into account projected rises in obesity. A recent report published by the World Health Organization attempts to produce comparable country-level estimates of obesity, as assessed by BMI, and to examine country-level trends in BMI (WHO Global Infobase Team 2005). Estimates for current and projected BMI (until 2010) are shown in Fig. 9.2. Average BMI, and thus levels of obesity, is expected to rise in all these high income countries. Since BMI is an important risk factor for coronary heart disease, stroke and several types of reproductive cancers (for example, post-menopausal breast cancer and

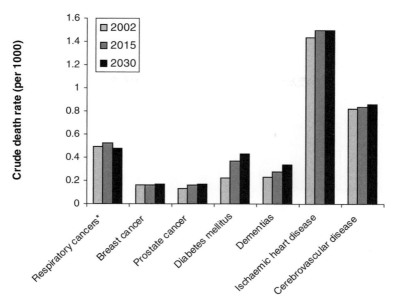

Fig. 9.1. Projected changes in the crude death rates for several western diseases in high income countries until 2030. Only the projections for diabetes mellitus take projected increases in body mass index into account. Much of the projected increase is attributable to the ageing of the population, which means that relatively more people will die of diseases that become more common with age. Data from Mathers and Loncar (2006). *Cancers of the trachea, bronchus and lung.

endometrial cancer), and appears also to be a risk factor for dementia, it is likely that projections that do not take an increase in BMI into account underestimate future rates of these diseases.

Other areas of the world

In less affluent areas of the world, death rates and DALYs attributable to non-communicable diseases are expected to rise more dramatically, even without taking rising obesity into account (Mathers and Loncar 2006). Much of this trend is accounted for by the fact that the increase in the proportion of older people will be even more rapid here than in affluent countries. Rapid trends towards socio-economic development and urbanisation are also expected and contribute to the projected rise in the burden of non-communicable disease. As with high-income countries, BMI is expected to rise over the next few years (Fig. 9.3) and this is likely to mean an even more dramatic increase in the mortality and disease burden attributable to these conditions.

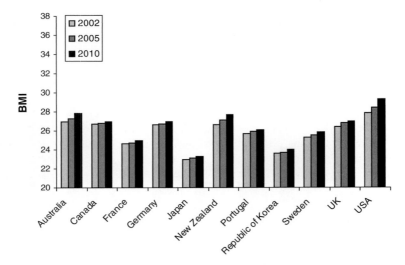

Fig. 9.2. World Health Organization estimates for current and projected BMI for selected high income countries until 2010 (WHO Global Infobase Team 2005). These projections suggest that the increases in crude death rates for all diseases shown in Fig. 9.1 are likely to be underestimates (except for the projections for type 2 diabetes, which take projected increases in BMI into account), since obesity increases the risk of these diseases, other than respiratory cancers.

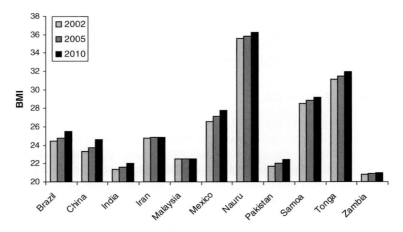

Fig. 9.3. World Health Organization estimates for current and projected BMI for selected low and middle income countries until 2010 (WHO Global Infobase Team 2005).

Unfortunately, prevalence data for diseases are less common and reliable in the developing world, so many figures have to be treated with caution. There is most information on BMI, blood pressure and prevalence of diabetes (assessed from fasting glucose or by oral glucose tolerance test), since they are relatively easily assessed in large-scale surveys.

In the mainly middle-income countries of Latin America and the Caribbean, levels of obesity are already high. In many of these countries BMI levels are similar to those seen in the US, and rates of adult obesity rose during the last two decades of the twentieth century (Martorell *et al.* 1998). Rates of many western diseases associated with obesity are thus also high in this part of the world. For example, surveys have shown that the high prevalence of abdominal obesity in Mexicans is associated with very high prevalences of diabetes and of hypertension, e.g. 33% of men and 26% of women aged 20–69 years had hypertension (Sánchez-Castillo *et al.* 2005).

In Asia, by contrast, the prevalence of obesity remains low compared to that in western countries (see Figs. 9.2 and 9.3). Even so, there have been dramatic recent increases in obesity. For example, in China there was 28 times as much overweight and obesity in children aged 7–18 years in 2000 compared to 1985 (Wu 2006). Not surprisingly, rates of type 2 diabetes have also been rising, again from a low baseline. China, Korea, Indonesia, Thailand, India, Singapore and Taiwan all had greater increases in the prevalence of type 2 diabetes over the last quarter of the twentieth century than did the USA (Yoon *et al.* 2006). In contrast to obesity, the current prevalence of type 2 diabetes across much of urban Asia is as high as that in the USA (Yoon *et al.* 2006). The rate of increase in type 2 diabetes, together with the rate of population growth, suggests that, by 2030, there will be more diabetics in India than in all the established market economies combined (Wild *et al.* 2004). There has also been an increase in the prevalence of hypertension and in hypercholesterolaemia in Asia, indicating increased cardiovascular disease risk (Wang *et al.* 2007).

It is clear that rapid socio-economic change in Asia is the main driver of the increases in obesity, type 2 diabetes and cardiovascular disease risk. There has been a large-scale shift from energy-intensive primary rural industry to occupations in services and manufacturing (Choi *et al.* 2006). In South Korea, over the last few decades of the twentieth century, there was also marked urbanisation; 28% of the population lived in urban areas in 1960, 82% in 1996 (Kim *et al.* 2000). In China there has been a reduction in manual labour, with, for example, large numbers of farmers leaving the land to work and live in urban areas, most of them in occupations that pay better and need less energy expenditure than farm work (Wang *et al.* 2007).

In South Korea and elsewhere, these changes have been associated with dramatic increases in private car ownership (293 000 in South Korea in 1983,

8 588 000 in 2001) and television watching (Choi *et al.* 2006). The 'nutrition transition' has been marked, with an increase in the consumption of animal protein and fat. In China, energy intake from fat increased from 22% to 30% between 1992 and 2002, while, for urban residents in 2002, fat contributed 35% of energy intake, more than the World Health Organization's suggested upper limit of 30% (Wang *et al.* 2007). Popkin and colleagues have shown how urbanisation in lower income countries is, in general, associated with a trend towards the consumption of food higher in fat, more animal produce, more sugar and more processed foods (Popkin 1999). The increasing availability of cheap vegetable oils is an important reason for increased fat consumption in such circumstances (Popkin and Gordon-Larsen 2004). Thus reduced physical activity, combined with detrimental changes associated with the nutrition transition, has resulted in increased obesity and higher rates of associated diseases, especially type 2 diabetes.

In general, people in Asia tend to have a higher proportion of body fat for a given level of BMI than do people of European origin (Yajnik *et al.* 2002; Yoon *et al.* 2006). While first noted in people of South Asian origin, there is now good evidence that Hong Kong Chinese, Indonesians, Singaporeans, urban Thai, young Japanese and Malays also have higher levels of body fat at a given BMI compared to people of European origin. In addition, fat patterning tends to favour abdominal obesity (Yoon *et al.* 2006). These trends are thought to account for the relatively high prevalence of type 2 diabetes observed in conjunction with relatively low levels of obesity, as assessed by BMI. There is, however, variation in body composition across populations and northern (more rural) Chinese and rural Thai appear to have similar values to Europeans (WHO Expert Consultation 2004), perhaps because of higher levels of muscle mass associated with physical activity. There has been some debate about whether lower BMI cut-off points for overweight and obesity should be adopted for Asian than for European populations (WHO Expert Consultation 2004; James 2005). At present, the World Health Organization has recommended that no new cut-off points should be introduced (WHO Expert Consultation 2004).

Why should Asian populations show body composition different from that typically seen in people of European origin? To answer this question, the arguments that were first offered to explain very high rates of type 2 diabetes in Pacific Islanders, Native Americans, Australian Aborigines and South Asians (Chapter 4) are now being extended to these populations. Thus the 'thrifty genotype' argument has been invoked (Yoon *et al.* 2006), but the effect of a low birth weight and a poorer early environment in general has been given more prominence (Pollard *et al.* in press). For example, Choi *et al.* (2006) speculate that thrifty phenotype effects may account for some of the rise in type 2 diabetes seen in Korea. They note that Korean people experienced malnutrition during the Japanese occupation (1909–1945) and the Korean War (1950–1953) and that babies born during those periods might be particularly susceptible to diabetes as a result of the relatively poor environment they experienced in early life. Other Asian populations also experienced much lower levels of energy intake in the recent past than they do now. For example, famine and chronic malnutrition caused the deaths of millions of people in China in the 1950s (Wu 2006). In many other parts of Asia, the contrast between the environment 50 years ago and today is less extreme, but is nevertheless much larger than in most western countries. As noted in Chapter 4, the effects of such a rapid transition are unfortunately not likely to be restricted to the generation that has lived through the changes (Fall 2001; Kuzawa 2005).

There have also been large changes in patterns of cancer incidence and mortality in China in recent years. The total number of cancer cases was projected to increase during the first 5 years of the twenty-first century, partly, but not entirely, due to rapid population growth and ageing (Yang *et al.* 2005). The biggest increases in incidence up to 2005 were predicted to be in lung cancer in both men and women, and breast cancer in women. The increase in lung cancer is strongly tied to large increases in tobacco smoking in China. With continuing increases in rates of smoking there and in other non-western countries, the rate of smoking-related deaths is likely to increase dramatically over the next few decades (West 2006). Yang *et al.* (2005), in line with the evidence reviewed in Chapter 5, blame westernisation of life-styles for the increase in breast cancer rates, specifically noting changes in diet, reduction in physical activity and rises in obesity. They also note the increasingly late age of women at first childbirth and low fertility, changes which are driven by China's fertility policies in addition to processes associated with westernisation.

In Africa, the poorest continent, the situation is different again. Currently the prevalence of obesity and related diseases is low, but urbanisation is proceeding at a particularly rapid rate. Large-scale surveys of the causes of morbidity and mortality are generally less well developed than in other parts of the world, but available data suggest that westernisation and urbanisation appear to be having a marked effect on the prevalence of obesity, type 2 diabetes and cardiovascular disease, particularly on hypertension and stroke.

Reported values for the prevalence of diabetes in rural areas of sub-Saharan Africa have been around 1% (Aspray *et al.* 2000), but rates are much higher amongst some urban populations. For example, data comparing adults from a village and urban area in Tanzania showed that urban residents had a significantly higher BMI and waist circumference than rural residents. Not surprisingly, they also had a significantly greater prevalence of diabetes, with age-adjusted rates of around 6% in the urban area, compared to 1%–2% in the rural area (Aspray *et al.* 2000). The projected percentage change in the number of people with diabetes between 2000 and 2030 is higher for Africa than it is even for India or China (Wild *et al.* 2004).

Similarly, the overall prevalence of hypertension in Africa appears, from limited data, to be currently relatively low, with around 10%–15% of the population affected (Cooper *et al.* 2003). However, urban populations in Africa show significantly higher mean systolic blood pressures than their rural counterparts (WHO Global Infobase Team 2005). A review of published data found that the prevalence of hypertension was around 3% in rural areas but greater than 30% in some urban settings (Mufunda *et al.* 2006). For example, in Ghana, rates of hypertension were higher in the regional capital than in rural areas of the Ashanti region (Agyemang 2006). Statistically the urban–rural difference in blood pressure was not totally explained by differences in BMI and smoking. Agyemang suggests that increasing stress, and a decline of the traditional social support systems due to the adoption of urban and western lifestyles, may contribute to high blood pressure levels. There are concerns that Africa will replace Europe as the region with the highest blood pressure levels in the world (WHO Global Infobase Team 2005) and since uncontrolled hypertension is an important risk factor for stroke, and also increases risk of heart failure and kidney failure, the likely impact on morbidity and mortality is large.

There is much less information on body composition (that is, on the relative proportion of body fat) in African than in Asian populations. The evidence available from one small study suggests that urban Africans from Ghana, unlike Asians, may not have proportionately more body fat than is observed in European populations (Osei *et al.* 1997). However, more evidence is needed on body composition in Africa.

In many parts of Asia and in sub-Saharan Africa, general awareness, treatment and control of hypertension are low, at least partly because many people cannot afford to pay for anti-hypertensive medications (Cooper *et al.* 2003; Agyemang 2006; Wang *et al.* 2007). Similarly, awareness and effective control of type 2 diabetes is often at a low level (Beran and Yudkin 2006). Health systems in these parts of the world are currently focused on the problems of infectious disease, and, in some cases, of undernutrition and malnutrition. The new double burden of non-communicable and infectious diseases, particularly that caused by HIV/AIDS in sub-Saharan Africa, places a huge financial burden on populations and health systems, consuming scarce resources (Tesfaye *et al.* 2007).

Allergic diseases have also extended to the urban centres of poorer countries in recent years (ISAAC Steering Committee 1998). Rates of asthma in Brazil, Costa Rica and Peru rival those seen in the UK and USA (ISAAC Steering Committee 1998). In one of the poorest countries in the world,

Ethiopia, a comparison of a remote rural area and a fairly small and unwesternised town showed that, while wheeze, asthma and atopic dermatitis were less common in both places than in the UK, they were more common in the town than in the rural area (Yemaneberhan *et al.* 1997; Yemaneberhan *et al.* 2004). This difference is probably explained by increased exposure to house dust mite (in houses built using modern housing construction methods with, for example, wooden floors) and perhaps by reduced endoparasite exposure in the town in comparison to the rural areas (Scrivener *et al.* 2001) (see Chapter 7). Thus, increasing urbanisation is likely to be associated with a rising prevalence of wheeze, asthma and atopic dermatitis. As with other western diseases, asthma is not well controlled in many developing countries (Zainudin *et al.* 2005).

To summarise, the prospects for rising rates of classic western diseases worldwide are very worrying. In the west rates of expected increase are low when increases in obesity are ignored, but it is clear that increasing rates of obesity will have a very adverse effect in the west. In poorer parts of the world, including Eastern Europe which is not considered in detail here, the rise in the impact of western diseases will be faster and more strongly felt, given the difficulties they create for already stretched health services.

Prevention of western diseases: an evolutionary perspective

Clearly, the need to act to stem the tide of western diseases is urgent, particularly in poorer parts of the world where the biggest rises in the numbers of people with western diseases are expected and where health systems are already struggling. Given current trends, the task is an enormous one. It is not my intention here to discuss all the possible strategies or the likely barriers to success; instead, I will focus on what an evolutionary perspective has to add to the debate about the way forward. Unfortunately, the evolutionary perspective is less useful here than it is in explaining the rise of western diseases. At its crudest, the obvious strategy arising from this perspective is an immediate return to life as hunter–gatherers. Such a strategy was in fact tried with a group of Australian Aborigines. O'Dea (1984) accompanied the group, some of whom were diabetic, while they lived for 7 weeks as hunter–gatherers in their traditional country in north-western Australia. The participants lost an average of 8 kg over the 7 weeks. Oral glucose tolerance tests, a standard diagnostic tool for diagnosing diabetes, showed that, on swallowing 75 g of glucose, participants had a smaller rise in glucose levels after living as hunter–gatherers than at baseline, indicating an improvement in their metabolic health (Fig. 9.4). This group retained many of the skills of

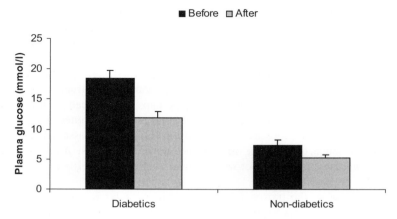

Fig. 9.4. Plasma glucose levels 2 hours after participants had swallowed 75 g glucose in ten diabetic and four non-diabetic Australian Aborigines, before and after 7 weeks of living as hunter–gatherers in their traditional country in north-western Australia. Bars show standard errors. Standard clinical interpretation is that a glucose level above 11.1 mmol/l indicates diabetes, a level of 7.8–11.1 mmol/l indicates impaired glucose tolerance, and a level below 7.8 mmol/l is normal. Data from O'Dea (1984).

their traditional hunter–gatherer way of life. The vast majority of the human population does not have such skills and would not choose this option if it were available. Fortunately, more realistic versions of what Eaton and Eaton (1988b) have called 'the Palaeolithic Prescription' are also available.

One important general issue raised by an evolutionary perspective is the ability it gives us to question our assumption that the human body in affluent western societies is normal. While western biomedicine has generally taken this approach, a closer look at cross-cultural and likely past human biology shows that it is often invalid. Thus, we should not necessarily reject interventions, including the use of pharmaceutical drugs, aimed at preventing western diseases on the basis that such strategies are not 'natural'. From an evolutionary perspective, there is nothing 'natural' about much of what westerners consume on a daily basis. Medicines may act on human biology to make it more normal in evolutionary terms.

Obesity, type 2 diabetes and cardiovascular disease

The importance of increased obesity as a risk factor for western diseases is clear and there have been concerted attempts to develop interventions to help the obese, or those diagnosed with, or at risk of, type 2 diabetes or cardio-vascular disease, to lose weight. There has been some success with strategies that focus on changes in diet and exercise. A meta-analysis of medically led weight loss programmes showed that short-term programmes based on diet alone, or diet and exercise together, typically cause a weight loss of around 11 kg, of which about 70% is sustained after 1 year (Miller *et al.* 1997). However, most of these studies have been conducted on middle-aged adults and it is not clear how far these results extend to other groups. Attempts to follow weight-loss diets advertised in the media tend to be much less successful, often because they rely on very low energy intakes that are not sustainable (Swinburn and Egger 2004). In children, interventions that aim to prevent weight gain, and others designed to limit sedentary behaviours such as watching television, have had some success, slowing age-related increases in BMI (DeMattia *et al.*2006).

Interventions to promote weight loss are much rarer in poorer countries. One randomised controlled trial of diet and/or exercise in men and women with impaired glucose tolerance living in Da Qing, China found that those who received advice about diet or exercise were less likely to develop type 2 diabetes than the control group (Pan *et al.* 1997), showing that this kind of intervention can be effective in poorer parts of the world. The suggestion has also been made that the maintenance of at least some traditional activities in westernising populations is likely to have protective effects on health. For example, Samoans who participate in farm work, which requires physical activity and also increases access to fruit and vegetables, had lower BMIs than those who did not (Keighley *et al.* 2007).

Programmes aimed at the overweight and obese, particularly those aimed at children, can raise ethical problems. There is now considerable stigma attached to obesity, with a widespread belief in western societies that obese people lack the discipline needed to maintain a normal weight (Friedman 2004). This effect may be a useful motivation to people to avoid weight gain (Swinburn and Egger 2004), but also has significant adverse consequences for the obese or even for those who are of normal weight but who perceive themselves to be obese. There are concerns about the rising prevalence of some eating disorders, such as bulimia nervosa, which may be caused partly by pressures, particularly on girls and young women, to be thin (Hudson *et al.* 2007). Such concerns can cause problems for surveys of levels of obesity and raise difficult problems for the handling of programmes designed to reduce levels of obesity, particularly in children. For example, in the new national survey of BMI in young children undertaken in 2005 and 2006 in England and Wales there were many anecdotal reports of selection bias in the sample, with overweight and obese children reported as being more likely to opt out of being measured (Crowther *et al.* 2007). In contrast, in some poorer countries being obese is valued as a sign of prosperity, raising rather different challenges for those aiming to reduce levels of obesity. For example, in Tonga, where levels of obesity are very high, people express a preference for larger body sizes (Craig *et al.* 1999).

Approaches that target what is termed the 'obesogenic' environment, rather than those that target individuals, have recently attracted growing interest, partly because they circumvent problems associated with efforts to reduce obesity by recommending that individuals change their behaviour. This kind of approach is more in line with the insights offered by an evolutionary perspective. Humans living as hunter–gatherers, and nearly all humans until the twentieth century, were thin because of the environment they lived in, not because they were able to control their appetites.

Obesogenic environments are those that promote high dietary energy intake and sedentary behaviour. They may have these effects by virtue of physical design, but the sociocultural rules that govern the environment are also important (Lake and Townshend 2006). Obesogenic environments make the unhealthy choices the easy, default choices (Swinburn and Egger 2004). In terms of the physical environment, there has been increasing concern to develop pedestrian-friendly neighbourhoods that encourage people to walk to local shops and services rather than use a car. To achieve this, population densities need to be high enough to support shops and services and the streets need to be safe for pedestrians (Lake and Townshend 2006). Residential density is currently particularly low in the USA, especially in the large suburban areas. Suburbs do not have sufficient population density and have been linked with obesity in a number of studies in the USA and Australia (Ewing *et al.* 2003; Lake and Townshend 2006). A study in San Diego reported that people in more walkable neighbourhoods on average walked 70 minutes more per week than did those in less walkable neighbourhoods (Saelens *et al.* 2003). Heavy marketing of energy dense foods, particularly to children, is a probably also a risk factor for obesity, and needs to be addressed in efforts to make the environment less obesogenic (Lake and Townshend 2006).

Gage (2005) asks ' . . . why cannot humans 'build' environments that are on the whole compatible with human biology?'. Given human ingenuity, he implies this ought to be possible. However, attempts to create environments that are compatible with human biology and not obesogenic are at a very early stage, and, given the inertia of both the physical and sociocultural environment, change is likely to be slow. Such change is particularly difficult in the newly developing urban environments of poorer countries, where only limited planning for public health is affordable and is likely to focus on the prevention of infectious diseases (for example, by providing clean water supplies) rather than on the prevention of non-communicable diseases.

Efforts to reduce the prevalence of cardiovascular disease rely partly on the reduction of obesity, but also target the fat content of the diet, fruit and vegetable consumption, and smoking. As noted in Chapter 4, both dietary fat content and smoking have declined in many western countries in the last few decades, presumably as a consequence of increased public understanding of their importance for health. For example, a dramatic reduction in mortality rates from coronary heart disease in Ireland was strongly related to declines in smoking and, particularly, to a fall in serum cholesterol levels which is itself probably related to progressively healthier eating patterns among the general Irish population (Bennett *et al.* 2006). However, adverse trends in obesity, diabetes and physical inactivity cancelled out over half the benefit associated with improved cholesterol levels between 1985 and 2000.

Efforts to prevent cardiovascular disease are far less developed in poorer countries. One positive example, though, is Mauritius, where an improvement in the serum lipid profile was reported in association with a non-communicable disease intervention programme, even though levels of obesity rose (Dowse *et al.* 1995). This was probably due to the fact that the Mauritian Ministry of Health ordered a change in the formulation of the most commonly used cooking oil from mainly palm oil to mainly soybean oil, rather than to individual decisions to make healthy food choices (Uusitalo *et al.* 1996). Murray *et al.* (2003) also highlight the advantages of non-personal health interventions, including government action to stimulate a reduction in the salt content of foods, as ways to reduce cardiovascular disease risk.

As the importance of a relatively poor early environment for later risk of type 2 diabetes and cardiovascular disease is increasingly recognised, researchers have started to address the possibility of reducing the rates of these diseases by improving early growth and development. As Fall (2001) suggests, the potential to prevent type 2 diabetes and cardiovascular disease can be added to a long list of reasons to recommend the promotion of healthy foetal and infant growth in poorer countries. Feeding pregnant women and their young children more and better food is likely to help. Other interventions during pregnancy can also increase birthweight, particularly smoking cessation and anti-malarial prophylaxis where malaria is endemic (James 2002). However, as we saw in Chapter 4, there is evidence for intergenerational transmission leading to phenotypic inertia in size and growth. This is predicted by evolutionary theory, since such intergenerational transfer provides the developing foetus with long-term cues about the environment, not just with information about the environment during the 9 months of its gestation (Kuzawa 2005). At its simplest, maternal height and weight predicts the birthweight of her offspring, so that small mothers generally have small babies. This implies that prevention should not just focus on the pregnant woman and growing infant and young child. Intervention to improve the growth of girls is needed, so that they become larger (but not obese!) women who have larger babies.

Reproductive cancers

Given clear links between lifetime exposure to endogenous reproductive steroids and increased risk of reproductive cancers in women, and the fact that we can be sure that the high levels of reproductive steroids seen in western women are evolutionarily novel (Chapter 5), intervention to reduce lifetime exposure to these hormones is an obvious preventive strategy. Several different approaches to reduce lifetime exposure to gonadal hormones have been suggested. Endogenous oestrogen protects against other diseases, such as osteoporosis, so the costs and benefits of the different approaches need to be carefully considered.

The first approach is behavioural intervention, designed to affect behaviours associated with a western lifestyle and high levels of gonadal hormones. Given the associations of energy balance and energy expenditure with ovarian function discussed in Chapter 5, it is clear that changes in diet and/or increases in physical activity levels would be useful. We have good evidence that, when a woman of reproductive age loses weight, her gonadal steroid levels are lowered in the ongoing menstrual cycle, and usually also in the following cycle (Lager and Ellison 1990; Jasieńska *et al.* 2000). So, in relation to diet, a reduction in energy intake, so as to help reverse the trend for western women to be in positive energy balance over much of their lives, would clearly be helpful. Reducing saturated fat levels in the diet has also been shown to reduce oestrogen levels (Wu *et al.* 1999) and there is some evidence that, in post-menopausal women who have had breast cancer, eating a low-fat diet can increase relapse-free survival (Chlebowski *et al.* 2006). Incorporation of more plant-derived oestrogens, for example, from soy products, into the diet has also been considered, but, given the complex and poorly understood ways in which exogenous oestrogens affect oestrogenic activity in the body, it is not clear that such changes in the diet of western women would have beneficial effects (Ellison 1999).

An increase in the levels of energy expenditure, even when not accompanied by negative energy balance, also causes suppression of progesterone and oestradiol levels (Lager and Ellison 1990; Jasieńska *et al.* 2000; Jasieńska *et al.* 2006b). Women who engaged in regular physical activity during their reproductive years have been shown to have a lower risk of breast cancer (Frisch *et al.* 1985; Bernstein *et al.* 1994). Apart from its effects on ovarian function, exercise also has the potential to improve insulin resistance and levels of related hormones and growth factors, changes that are expected to reduce cancer risk. Physical activity also has benefits in relation to diseases associated with low oestrogen levels in western societies including, most importantly, osteoporosis and cardiovascular disease. Exercise also, of course, reduces the risk of type 2 diabetes and cardiovascular disease and there is preliminary evidence that it may even be an effective treatment for depression (Lett *et al.* 2004). Thus, increasing physical activity levels appears to be a promising approach to reduce the incidence of reproductive cancers, with only positive side effects (Ellison 1999).

While a western lifestyle is also associated with high levels of gonadal steroids in men, the proximate determinants of testosterone levels in men are not the same as the proximate determinants of progesterone and oestradiol levels in women (see pp. 88–93 and 96–97). Moderate exercise does not appear to have a noticeable effect on salivary testosterone levels in men (Ellison 1999; Bribiescas 2001b). Thus, it is not clear that increased activity levels in men would be expected to protect against prostate cancer, and findings from studies of this association are inconsistent (Saxton 2006). There is some evidence that combining very low fat diets with regular physical activity can induce significant changes in IGF-1 and its binding protein in men that could protect against the development or progression of prostate cancer (Saxton 2006), but attaining the low level of fat required may not be a realistic goal for most men.

Attempts to reduce exposure to oestrogen by using drugs have been made. Aromatase inhibitors suppress plasma oestrogen levels by inhibiting or inactivating aromatase, the enzyme that catalyses the conversion of androgens to oestrogens in the ovaries and in subcutaneous fat (the most important source of oestrogens in post-menopausal women). The other major drug-based approach to breast cancer prevention in post-menopausal women is to give selective oestrogen receptor modulators that change the ways in which oestrogen acts on the body. There has been concern about side effects associated with these drugs. Tamoxifen, the best known of them, increases the risk of thromboembolism (obstruction of a blood vessel by a blood clot formed elsewhere in the body) and endometrial cancer, but raloxifene, a newer drug of this type, has far fewer side effects (Vogel *et al.* 2006). However, so far, only women at high risk of breast cancer (for example, because close relatives have had the disease) have been prescribed drugs in order to prevent breast cancer. The induction of an early menopause either surgically, by removing the ovaries, or chemically, would also protect against reproductive cancers, but is unlikely to prove an appealing option to most women.

Late age at first pregnancy, or no full-term pregnancy, are important risk factors for breast cancer. Clearly, the potential to modify women's choices regarding pregnancy because of the effects on breast cancer risk is limited (Ellison 1999). Instead, it has been suggested that pregnancy might be pharmacologically simulated in adolescent women, thus achieving the effect of natural pregnancy in relation to terminal differentiation of duct tissue in the breast. This approach aims to mimic the effect of pregnancy at an early age, typical of women in evolutionary history (Eaton *et al.* 1994). Work to establish its feasibility continues. It has been shown that giving both natural and synthetic oestrogens in combination with progestins to rodents at doses that simulate levels in late pregnancy can prevent breast cancer (Rajkumar *et al.* 2004; Tonetti 2004). However, there have been no trials of this approach in humans and there are clear ethical problems in conducting this kind of experimental work with humans, particularly adolescents.

Breastfeeding has a suppressive effect on the ovaries and so increasing the proportion of women successfully initiating breastfeeding, and, even more importantly, extending the duration of breastfeeding could protect women from reproductive cancers. An analysis of pooled data from 47 studies showed that the risk of breast cancer was lower for women who had breastfed than for women who had not, and that the risk decreased with increasing duration of breastfeeding (Beral *et al.* 2002).

Other benefits of breastfeeding

An increase in breastfeeding could have numerous other health benefits. Given known associations of breastfeeding with disease risks (see pp. 105–110), an increase in breastfeeding is likely to protect the mother against weight gain following pregnancy and against subsequent development of insulin resistance. Having been breastfed is likely to protect an individual against gastrointestinal and respiratory illness during infancy, disorders of immune regulation (including allergies), obesity, insulin resistance and type 2 diabetes, developing a harmful serum lipid profile and high blood pressure. It may also protect him or her from type 1 diabetes and sudden infant death syndrome (because of the association of breastfeeding with mother–infant co-sleeping) and lead to improved cognitive development. Increased breast-feeding is one area in which women can mimic the behaviour of our ancestors, although this will also involve the removal of cultural and socio-economic barriers such as restricted maternity leave.

Allergy and autoimmune disease

The reformulated hygiene hypothesis suggests that the developing human immune system expects to encounter bacteria and helminths that would have been common during most of human evolutionary history. Suggestions have therefore been made that the development of allergy could be inhibited by exposing infants to probiotics, dietary supplements containing bacteria that are thought to be beneficial for health. Trials of probiotics have begun. Two studies have been conducted in which probiotics were given to pregnant mothers with a family history of atopic disease before their expected delivery, and then to the newborn babies for 6 months after delivery, while placebo was given to another group. Both found that the group taking probiotics had a reduced risk of atopic eczema (the most common allergic disease in infants) compared with the group taking placebo (Kalliomäki *et al.* 2001; Kukkonen *et al.* 2007). However, another found that, when newborns of women with allergy received either *Lactobacillus acidophilus* or placebo daily for the first 6 months of life, there seemed to be no effect on the prevalence of atopic eczema and, worse, there was evidence of an increase in allergic sensitisation (Taylor *et al.* 2007). Unfortunately, these studies focus on the effects of children at high risk of allergy based on the familial prevalence of allergic diseases, and little is known about the effects of probiotics on the rest of the population. Nevertheless, it is clear that there is potential for probiotics to have a role in preventing allergic disease.

Another approach currently being investigated is based on injecting children with *Mycobacterium vaccae*, one of the many Mycobacteria commonly found in soil and untreated water. Clinical studies of its efficacy in either the therapy or prevention of allergic disease are at an early stage (Matricardi *et al.* 2003). A wider anti-inflammatory approach, which focuses on stimulating the newly discovered regulatory T-cells, thought to be important in preventing allergy (see p. 128–129), is also being considered (Matricardi *et al.* 2003). However, the full function of regulatory T-cells is not yet understood, and it is possible that they might impair the immune system's ability to deal with chronic infections or to act against cancerous cells (Akdis *et al.* 2005).

Mental health and associated disease outcomes

As with obesity, there are two main approaches to preventing the experience of depression or stress or other mental health problems. One is to act at the level of the individual to change his or her behaviour or cognitive style. A meta-analysis of such studies concluded that it is possible to reduce the incidence of new mental disorders, including depression, in this way (Cuijpers *et al.* 2005). These strategies have most often been applied to those with existing mental health problems or with chronic diseases, such as coronary heart disease or cancer, or at high risk of mental health problems for other reasons. Overall, there is evidence that they can increase quality of life and improve clinical outcomes (Linden *et al.* 1996; Lett *et al.* 2004). However, the interventions used for patients with cardiovascular disease are intensive and not realistic on a large scale.

The other approach is to change the environment so that it is less likely to engender mental health problems. In relation to job-related stress, modifications of the occupational environment to give workers more control have been suggested. This approach has been most applied in Scandinavia (Orth-Gomer *et al.* 1994), but appears to have been largely ineffective (van der Klink *et al.* 2001). It is very difficult to intervene to increase job control, identified as a key determinant of job stress, without fundamentally changing the structure of work (Macleod and Davey Smith 2003). On a larger scale, attempts to make changes to the social structure of populations to improve the psychosocial experience of individuals seem unlikely.

Summary

Western diseases present a major threat to human well-being today. They are widespread and are expected to become even more common in the next few decades, particularly in poorer parts of the world. In more affluent countries huge resources are currently poured into efforts to prevent and treat these diseases, while in poorer parts of the world health services are still generally focused on infectious diseases. In the future, efforts to prevent western diseases will become increasingly important worldwide.

It is clear that obesity is a major risk factor for many western diseases, and a secondary risk factor for others. Furthermore, many perceived threats to health, such as the menopause and stress, are probably greatly exacerbated by life in an environment that makes it easy to consume excess energy and to be physically inactive. It seems likely, therefore, that persuading people to change behaviours that affect their risk of becoming obese, and perhaps even more importantly, modifying the environment to make it less obesogenic, will be the most important preventive strategies to pursue in the coming decades. More specifically, an evolutionary perspective suggests two key targets for health promotion strategies. These are physical activity levels and breastfeeding behaviour. Increases in both would have multiple benefits for health, over and above positive effects on obesity levels. Implementation of such strategies will require tremendous effort and input from governments and agencies concerned with health.

An evolutionary perspective provides important insights into the causes of western diseases. It may also help in the implementation of preventive strategies, by indicating the key features of our environment that we need to change, and by helping people to understand that their bodies are not adapted by evolution to the modern western environment.

References

Abate, N., Carulli, L., Cabo-Chan, A., Chandalia, M., Snell, P. G. and Grundy, S. M. (2003). Genetic polymorphism PC-1 K121Q and ethnic susceptibility to insulin resistance. *Journal of Clinical Endocrinology and Metabolism*, **88**, 5927–34.

Adler, A. I., Stratton, I. M., Neil, A. W. *et al.* on behalf of the UK Prospective Diabetes Study Group (2000). Association of systolic blood pressure with macrovascular and microvascular complications of type 2 diabetes (UKPDS 36): prospective observational study. *British Medical Journal*, **321**, 412–19.

Adlerberth, I., Lindberg, E., Aberg, N. *et al.* (2005). Reduced enterobacterial and increased staphylococcal colonization of the infantile bowel: an effect of hygienic lifestyle? *Pediatric Research*, **59**, 96–101.

Adlercreutz, H. (2002). Phyto-estrogens and cancer. *The Lancet Oncology*, **3**, 364–373.

Agyemang, C. (2006). Rural and urban differences in blood pressure and hypertension in Ghana, West Africa. *Public Health*, **120**, 525–33.

Aiello, L. C. and Wells, J. C. K. (2002). Energetics and the evolution of the genus *Homo*. *Annual Review of Anthropology*, **31**, 323–38.

Aiello, L. C. and Wheeler, P. (1995). The expensive tissue hypothesis: the brain and digestive system in human and primate evolution. *Current Anthropology*, **36**, 199–221.

Akdis, M., Blaser, K. and Akdis, C. A. (2005). T regulatory cells in allergy: novel concepts in the pathogenesis, prevention, and treatment of allergic diseases. *Journal of Allergy and Clinical Immunology*, **116**, 961–9.

Alavanja, M. C. R., Hoppin, J. A. and Kamel, F. (2004). Health effects of chronic pesticide exposure: cancer and neurotoxicity. *Annual Review of Public Health*, **25**, 155–97.

Alberti, K. G. M. M., Zimmet, P. and Shaw, J. for the IDF Epidemiology Task Force Consensus Group (2005). The metabolic syndrome – a new worldwide definition. *The Lancet*, **366**, 1059–62.

Aligne, C. A., Auinger, P., Byrd, R. S. and Weitzman, M. (2000). Risk factors for pediatric asthma: contributions of poverty, race, and urban residence. *American Journal of Respiratory and Critical Care Medicine*, **162**, 873–7.

Allen, J. S. and Cheer, S. M. (1996). The non-thrifty genotype. *Current Anthropology*, **37**, 831–42.

Allen, N. B. and Badcock, P. B. T. (2003). The social risk hypothesis of depressed mood: evolutionary, psychosocial, and neurobiological processes. *Psychological Bulletin*, **129**, 887–913.

173

Allen, N. B. and Badcock, P. B. T. (2006). Darwinian models of depression: a review of evolutionary accounts of mood and mood disorders. *Progress in Neuro-Psychopharmacology and Biological Psychiatry*, **30**, 815–26.

Allen, N. E. and Key, T. J. (2000). The effects of diet on circulating sex hormone levels in men. *Nutrition Research Reviews*, **13**, 159–84.

Anand, S. S., Yusuf, S., Jacobs, R. *et al.* for the SHARE-AP Investigators (2001). Risk factors, atherosclerosis, and cardiovascular disease among Aboriginal people in Canada: the Study of Health Assessment and Risk Evaluation in Aboriginal Peoples (SHARE-AP). *The Lancet*, **358**, 1147–53.

Anderson, H. R. (1997). Air pollution and trends in asthma. In *The Rising Trends in Asthma*, ed. D. Chadwick and G. Cardew. New York: John Wiley and Sons, pp. 190–203.

Anderson, H. R., Ruggles, R., Strachan, D. P. *et al.* (2004). Trends in prevalence of symptoms of asthma, hay fever, and eczema in 12–14 year olds in the British Isles, 1995–2002: questionnaire survey. *British Medical Journal*, **328**, 1052–3.

Anderson, J. W., Johnstone, B. M. and Remley, D. T. (1999). Breast-feeding and cognitive development: a meta-analysis. *American Journal of Clinical Nutrition*, **70**, 525–35.

Anderson, M. (1988). *Population Change in North-Western Europe, 1750–1850*. Basingstoke, UK: Macmillan Education.

Apter, D. and Vihko, R. (1990). Endocrine determinants of fertility: serum androgen concentrations during follow-up of adolescents into the third decade of life. *Journal of Clinical Endocrinology and Metabolism*, **71**, 970–4.

Apter, D., Reinila, M. and Vihko, R. (1989). Some endocrine characteristics of early menarche, a risk factor for breast cancer, are preserved into adulthood. *International Journal of Cancer*, **44**, 783–7.

Arnett, D. K., Xiong, B., McGovern, P. G., Blackburn, H. and Luepker, R. V. (2000). Secular trends in dietary macronutrient intake in Minneapolis-St. Paul, Minnesota, 1980–1992. *American Journal of Epidemiology*, **152**, 868–73.

Asher, M. I., Montefort, S., Björkstén, B. *et al.* and the ISAAC Phase Three Study Group (2006). Worldwide time trends in the prevalence of symptoms of asthma, allergic rhinoconjunctivitis, and eczema in childhood: ISAAC Phases One and Three repeat multicountry cross-sectional surveys. *The Lancet*, **368**, 733–43.

Asia–Pacific Cohort Studies Collaboration (2005). Smoking, quitting, and the risk of cardiovascular disease among women and men in the Asia-Pacific region. *International Journal of Epidemiology*, **34**, 1036–45.

Aspray, T. J., Mugusi, F., Rashid, S. *et al.* for the Essential Non-Communicable Disease Health Intervention Project (2000). Rural and urban differences in diabetes prevalence in Tanzania: the role of obesity, physical inactivity and urban living. *Transactions of the Royal Society of Tropical Medicine and Hygiene*, **94**, 637–44.

Astrup, A. and Finer, N. (2000). Redefining type 2 diabetes: 'Diabesity' or 'Obesity dependent diabetes mellitus'? *Obesity Reviews*, **1**, 57–9.

Atwood, L. D., Heard-Costa, N. L., Cupples, L. A., Jaquish, C. E., Wilson, P. W. F. and D'Agostino, R. B. (2002). Genomewide linkage analysis of body mass index across 28 years of the Framingham Heart Study. *American Journal of Human Genetics*, **71**, 1044–50.

Baier, L. J. and Hanson, R. L. (2004). Genetic studies of the etiology of type 2 diabetes in Pima Indians. *Diabetes*, **53**, 1181–6.

Baker, P. T. (1984). Migration, genetics, and the degenerative diseases of South Pacific Islanders. In *Migration and Mobility: Biosocial Aspects of Human Movement*, ed. A. J. Boyce. London: Taylor and Francis, pp. 209–39.

Baker, P. T. (1988). Infectious disease. In *Human Biology: An Introduction to Human Evolution, Variation, Growth, and Adaptability*, ed. G. A. Harrison, J. M. Tanner, D. R. Pilbeam and P. T. Baker. Oxford: Oxford University Press, pp. 508–28.

Baker, P. T., Hanna, J. M. and Baker, T. S., eds. (1986). *The Changing Samoans: Behavior and Health in Transition*. New York: Oxford University Press.

Balen, A. (1999). Pathogenesis of polycystic ovary syndrome – the enigma unravels? *The Lancet*, **354**, 966–7.

Ball, H. L. (2006). Parent–infant bed-sharing behavior: effects of feeding type and presence of father. *Human Nature*, **17**, 301–18.

Ball, H. L. and Klingaman, K. (2008). Breastfeeding and mother–infant sleep proximity: implications for infant care. In *New Perspectives in Evolutionary Medicine*, ed. W. R. Trevathan, E. O. Smith and J. J. McKenna. New York: Oxford University Press, pp. 226–241.

Barker, D. J. P. (1994). *Mothers, Babies, and Disease in Later Life*. London: BMJ Publishing Group.

Barker, D. J. P. and Osmond, C. (1986). Infant mortality, childhood nutrition and ischaemic heart disease in England and Wales. *The Lancet*, **i**, 1077–81.

Barker, D. J. P., Osmond, C., Forsen, T. J., Kajantie, E. and Eriksson, J. G. (2005). Trajectories of growth among children who have coronary events as children. *New England Journal of Medicine*, **353**, 1802–09.

Barnes, K. C. (2006). Genetic epidemiology of health disparities in allergy and clinical immunology. *Journal of Allergy and Clinical Immunology*, **117**, 243–54.

Barnes, K. C., Armelagos, G. J. and Morreale, S. C. (1999). Darwinian medicine and the emergence of allergy. In *Evolutionary Medicine*, ed. W. R. Trevathan, E. O. Smith and J. J. McKenna. New York: Oxford University Press, pp. 209–43.

Barrett, R., Kuzawa, C. W., McDade, T. and Armelagos, G. J. (1998). Emerging and re-emerging infectious diseases: the third epidemiologic transition. *Annual Review of Anthropology*, **27**, 247–71.

Barroso, I. (2005). The genetics of type 2 diabetes. *Diabetic Medicine*, **22**, 517–35.

Baschetti, R. (1998). Diabetes epidemic in newly westernized populations: is it due to thrifty genes or to genetically unknown foods. *Journal of the Royal Society of Medicine*, **91**, 622–25.

Bassuk, S. S. and Manson, J. E. (2005). Epidemiological evidence for the role of physical activity in reducing risk of type 2 diabetes and cardiovascular disease. *Journal of Applied Physiology*, **99**, 1193–204.

Ben-Shlomo, Y. and Davey Smith, G. (1991). Deprivation in infancy or in adult life: which is more important for mortality risk? *The Lancet*, **337**, 530–4.

Bennett, K., Kabir, Z., Unal, B. *et al.* (2006). Explaining the recent decrease in coronary heart disease mortality rates in Ireland, 1985–2000. *Journal of Epidemiology and Community Health*, **60**, 322–7.

Bentley, G. R. (2000). Environmental pollutants and fertility. In *Infertility in the Modern World: Present and Future Prospects*, ed. G. R. Bentley and C. G. N. Mascie-Taylor. Cambridge: Cambridge University Press, pp. 85–152.

Bentley, G. R., Jasieńska, G. and Goldberg, T. (1993). Is the fertility of agriculturalists higher than that of nonagriculturalists? *Current Anthropology*, **34**, 778–85.

Bentley, G. R., Harrigan, A. M. and Ellison, P. T. (1998). Dietary composition and ovarian function among Lese horticulturalist women of the Ituri Forest, Democratic Republic of Congo. *European Journal of Clinical Nutrition*, **52**, 261–70.

Bentley, G. R., Paine, R. R. and Boldsen, J. L. (2001). Fertility changes with the prehistoric transition to agriculture. In *Reproductive Ecology and Human Evolution*, ed. P. Ellison. New York: Aldine de Gruyter, pp. 201–31.

Benyshek, D. C. and Watson, J. T. (2006). Exploring the thrifty genotype's food-shortage assumptions: a cross-cultural comparison of ethnographic accounts of food security among foraging and agricultural societies. *American Journal of Physical Anthropology*, **131**, 120–6.

Beral, V., Bull, D., Doll, R. *et al.* (1997). Breast cancer and hormone replacement therapy: collaborative reanalysis of data from 51 epidemiological studies of 52 705 women with breast cancer and 108 411 women without breast cancer. *The Lancet*, **350**, 1047–59.

Beral, V., Bull, D., Doll, R. *et al.* (2002). Breast cancer and breastfeeding: collaborative reanalysis of individual data from 47 epidemiological studies in 30 countries, including 50 302 women with breast cancer and 969 973 women without the disease. *The Lancet*, **360**, 187–95.

Beran, D. and Yudkin, J. S. (2006). Diabetes care in sub-Saharan Africa. *The Lancet*, **368**, 1689–95.

Bernstein, L. and Ross, R. K. (1993). Endogenous hormones and breast cancer risk. *Epidemiologic Reviews*, **15**, 48–65.

Bernstein, L., Yuan, J. M., Ross, R. K. *et al.* (1990). Serum hormone levels in premenopausal Chinese women in Shanghai and white women in Los Angeles: results from two breast cancer case-control studies. *Cancer Causes and Control*, **1**, 51–8.

Bernstein, L., Henderson, B. E., Hanisch, R., Sullivan-Halley, J. and Ross, R. K. (1994). Physical exercise and reduced risk of breast cancer in young women. *Journal of the National Cancer Institute*, **86**, 1403–08.

Berrington, A. (2004). Perpetual postponers? Women's, men's and couple's infertility intentions and subsequent fertility behaviour. *Population Trends*, **117**, 9–19.

Biesele, M. and Howell, N. (1981). "The old people give you life": aging among !Kung hunter–gatherers. In *Other Ways of Growing Old*, ed. P. Amoss and S. Harrell. Stanford: Stanford University Press, pp. 77–98.

Bilsborough, S. and Mann, N. (2006). A review of issues of dietary protein intake in humans. *International Journal of Sport Nutrition and Exercise Metabolism*, **16**, 129–52.

Bindon, J. R. and Baker, P. T. (1997). Bergmann's rule and the thrifty genotype. *American Journal of Physical Anthropology*, **104**, 201–10.

Bjerve, K. S., Brubakk, A. M., Fougner, K. H., Johnsen, H., Midthjell, K. and Vik, T. (1993). Omega-3 fatty acids: essential fatty acids with important biological effects, and serum phospholipid fatty acids as markers of dietary omega-3 fatty acid intake. *American Journal of Clinical Nutrition*, **57** (Suppl.), 801S–6S.

Björkstén, B., Naaber, P., Sepp, E. and Mikelsaar, M. (1999). The intestinal microbiota in allergic Estonian and Swedish 2-year-old children. *Clinical and Experimental Allergy*, **29**, 342–6.

Björntorp, P. and Rosmond, R. (2000). Obesity and cortisol. *Nutrition*, **16**, 924–36.

Blackley, R. (2001). The arrival of the Maoris in New Zealand, 1898. *Emerging Infectious Diseases*, **7**, 914.

Blair, P. S. and Ball, H. L. (2004). The prevalence and characteristics associated with parent–infant bed-sharing in England. *Archives of Disease in Childhood*, **89**, 1106–10.

Blumenthal, M. N., Langefeld, C. D., Beaty, T. H. *et al.* (2004). A genome-wide search for allergic response (atopy) genes in three ethnic groups: Collaborative Study on the Genetics of Asthma. *Human Genetics*, **114**, 157–64.

Blurton Jones, N. G., Hawkes, K. and O'Connell, J. F. (2002). Antiquity of postreproductive life: are there modern impacts on hunter–gatherer postreproductive life spans? *American Journal of Human Biology*, **14**, 184–205.

Bolling, K. (2006). *Infant Feeding Survey 2005: Early Results*. London: Information Centre, Government Statistical Service, p. 19.

Bonnet, M. H. and Arand, D. L. (1995). We are chronically sleep deprived. *Sleep*, **18**, 908–11.

Botha, J. L., Bray, F., Sankila, R. and Parkin, D. M. (2003). Breast cancer incidence and mortality trends in 16 European countries. *European Journal of Cancer*, **39**, 1718–29.

Bowlby, J. (1969). *Attachment and Loss*. New York: Basic Books.

Boyd, R. and Silk, J. B. (2006). *How Humans Evolved*. Los Angeles, University of California: University of California Press.

Boyden, S. V., ed. (1970). *The Impact of Civilization on the Biology of Man*. Canberra: Australian National University Press.

Boyden, S. V. (1987). *Western Civilization in Biological Perspective: Patterns in Biohistory*. Oxford: Oxford University Press.

Braman, S. S. (2006). The global burden of asthma. *Chest*, **130**, 4S–12S.

Braun, L. (2002). Race, ethnicity and health: can genetics explain disparities? *Perspectives in Biology and Medicine*, **45**, 159–74.

Bribiescas, R. G. (2001a). Reproductive ecology and life history of the human male. *Yearbook of Physical Anthropology*, **44**, 148–76.

Bribiescas, R. G. (2001b). Reproductive physiology of the human male: an evolutionary and life history perspective. In *Reproductive Ecology and Human Evolution*, ed. P. T. Ellison. New York: Aldine de Gruyter, pp. 107–35.

Briefel, R. R. and Johnson, C. L. (2004). Secular trends in dietary intake in the United States. *Annual Review of Nutrition*, **24**, 401–31.

Brock, J. (1993). *Native Plants of Northern Australia*. New Holland: Frenchs Forest, Reed.

Brown, D. E. (2006). Measuring hormonal variation in the sympathetic nervous system: catecholamines. In *Measuring Stress in Humans: A Practical Guide for the Field*, ed. G. H. Ice and G. D. James. Cambridge: Cambridge University Press, pp. 94–121.

Brown, E. S., Varghese, F. P. and McEwen, B. S. (2004). Association of depression with medical illness: does cortisol play a role? *Biological Psychiatry*, **55**, 1–9.

Brown, P. J. and Bentley-Condit, V. K. (1998). Culture, evolution, and obesity. In *Handbook of Obesity*, ed. G. A. Bray, C. Bouchard and W. P. T. James. New York: Marcel Dekker, pp. 143–55.

Brown, P. J. and Konner, M. (1987). An anthropological perspective on obesity. *Annals of the New York Academy of Sciences*, **499**, 29–46.

Brown, P. J. and Krick, S. V. (2001). The etiology of obesity: diet, television and the illusions of personal choice. In *Obesity, Growth and Development*, ed. F. E. Johnston and G. D. Foster. London: Smith-Gordon, pp. 111–28.

Brunner, E. and Marmot, M. (1999). Social organization, stress, and health. In *Social Determinants of Health*, ed. M. Marmot and R. G. Wilkinson. Oxford: Oxford University Press, pp. 17–43.

Burkitt, D. P. (1973). Some diseases characteristic of modern western civilization. *British Medical Journal*, **1**, 274–8.

Butte, N. F. (2001). The role of breastfeeding in obesity. *Pediatric Clinics of North America*, **48**, 189–98.

Cacioppo, J. T. and Hawkley, L. C. (2003). Social isolation and health, with an emphasis on underlying mechanisms. *Perspectives in Biology and Medicine*, **46**, S39–52.

Caldwell, J. C. (2001). Demographers and the study of mortality: scope, perspectives and theory. *Annals of the New York Academy of Sciences*, **954**, 19–34.

Calle, E. E. and Kaaks, R. (2004). Overweight, obesity and cancer: epidemiological evidence and proposed mechanisms. *Nature Reviews Cancer*, **4**, 579–91.

Calvani, M., Alessandri, C. and Bonci, E. (2002). Fever episodes in early life and the development of atopy in children with asthma. *European Respiratory Journal*, **20**, 391–6.

Camacho, T. C., Roberts, R. E., Lazarus, N. B., Kaplan, G. A. and Cohen, R. D. (1991). Physical activity and depression: evidence from the Alameda County Study. *American Journal of Epidemiology*, **134**, 220–31.

Carmichael, A. R. and Bates, T. (2004). Obesity and breast cancer: a review of the literature. *The Breast*, **13**, 85–92.

Carter, H. (1999). Urban origins: a review of theories. In *The Pre-Industrial Cities and Technology Reader*, ed. C. Chant. London: Routledge and The Open University, pp. 7–14.

Carulli, L., Rondinella, S., Lombardini, S., Canedi, I., Loria, P. and Carulli, N. (2005). Diabetes, genetics and ethnicity. *Alimentary Pharmacology and Therapeutics*, **22** (Suppl. 2), 16–19.

Catanese, D. M., Koetting O'Byrne, K. and Poston, W. S. C. (2001). The epidemiology of obesity in developed countries. In *Obesity, Growth and Development*, ed. F. Johnston and G. Foster. London: Smith-Gordon, pp. 69–90.

Cavalli-Sforza, L. L. and Feldman, M. W. (2003). The application of molecular genetic approaches to the study of human evolution. *Nature Genetics*, **33**, 266–75.

Cavallo, M. G., Fava, D., Monetini, L., Barone, F. and Pozzilli, P. (1996). Cell-mediated immune response to β casein in recent-onset insulin dependent diabetes: implications for disease pathogenesis. *The Lancet*, **348**, 926–8.

Chisholm, J. S. and Burbank, V. K. (2001). Evolution and inequality. *International Journal of Epidemiology*, **30**, 206–11.

Chlebowski, R. T., Blackburn, G. L., Thomson, C. A. *et al.* (2006). Dietary fat reduction and breast cancer outcome: interim efficacy results from the Women's Intervention Nutrition Study. *Journal of the National Cancer Institute*, **98**, 1767–76.

Chobanian, A. V. and Alexander, R. W. (1996). Exacerbation of atherosclerosis by hypertension. *Archives of Internal Medicine*, **156**, 1952–6.

Choi, Y. J., Cho, Y. M., Park, C. K. *et al.* (2006). Rapidly increasing diabetes-related mortality with socio-environmental changes in South Korea during the last two decades. *Diabetes Research and Clinical Practice*, **74**, 295–300.

Chow, C. K., Raju, P. K., Raju, R. *et al.* (2006). The prevalence and management of diabetes in rural India. *Diabetes Care*, **29**, 1717–18.

Cleave, T. L., Campbell, G. D. and Painter, N. S. (1969). *Diabetes, Coronary Thrombosis, and the Saccharine Disease*. Bristol: Wright.

Coe, K. and Steadman, L. B. (1995). The human breast and the ancestral reproductive cycle: a preliminary inquiry into breast cancer etiology. *Human Nature*, **6**, 197–220.

Cohen, M. N. (1989). *Health and the Rise of Civilization*. New Haven: Yale University Press.

Cohen, S. (2005). The Pittsburgh common cold studies: psychosocial predictors of susceptibility to respiratory infectious illness. *International Journal of Behavioral Medicine*, **12**, 123–31.

Cohen, S., Doyle, W. and Baum, A. (2006). Socioeconomic status is associated with stress hormones. *Psychosomatic Medicine*, **68**, 414–20.

Colagiuri, S., Colagiuri, R., Na'ati, S., Muimuiheata, S., Hussain, Z. and Palu, T. (2002). The prevalence of diabetes in the Kingdom of Tonga. *Diabetes Care*, **25**, 1378–83.

Cole, T. J., Bellizzi, M. C., Flegal, K. M. and Dietz, W. H. (2000). Establishing a standard definition for child overweight and obesity worldwide: international survey. *British Medical Journal*, **320**, 1240–3.

Colla, J., Buka, S., Harrington, D. and Murphy, J. M. (2006). Depression and modernization: a cross-cultural study of women. *Social Psychiatry and Psychiatric Epidemiology*, **41**, 271–9.

Collins, P. (2001). GPs and stress discourse: a preliminary report. *Anthropology in Action*, **8**, 36–44.

Cook, D. C. (1984). Subsistence and health in the lower Illinois Valley: osteological evidence. In *Palaeopathology at the Origins of Agriculture*, ed. M. N. Cohen and G. J. Armelagos. Orlando, Florida: Academic Press, pp. 235–69.

Cooke, A., Zaccone, P., Raine, T., Phillips, J. M. and Dunne, D. W. (2004). Infection and autoimmunity: are we winning the war, only to lose the peace? *Trends in Parasitology*, **20**, 316–21.

Cookson, W. O. and Moffatt, M. F. (1997). Asthma: an epidemic in the absence of infection? *Science*, **275**, 41–2.

Cooper, P. J. (2004). Intestinal worms and human allergy. *Parasite Immunology*, **26**, 455–67.

Cooper, R. S., Rotimi, C., Ataman, S. *et al.* (1997). The prevalence of hypertension in seven populations of West African origin. *American Journal of Public Health*, **87**, 160–8.

Cooper, R. S., Amoah, A. G. and Mensah, G. A. (2003). High blood pressure: the foundation for epidemic cardiovascular disease in African populations. *Ethnicity and Disease*, **13** (Suppl. 2), S48–52.

Cordain, L., Gotshall, R. W. and Eaton, S. B. (1997). Evolutionary aspects of exercise. *World Review of Nutrition and Dietetics*, **81**, 49–60.

Cordain, L., Brand Miller, J., Eaton, S. B., Mann, N., Holt, S. H. A. and Speth, J. D. (2000). Plant–animal subsistence ratios and macronutrient energy estimations in worldwide hunter–gatherer diets. *American Journal of Clinical Nutrition*, **71**, 682–92.

Cordain, L., Eaton, S. B., Brand Miller, J., Mann, N. and Hill, K. (2002). The paradoxical nature of hunter–gatherer diets: meat-based, yet non-atherogenic. *European Journal of Clinical Nutrition*, **56** (Suppl. 1), S42–52.

Cordain, L., Eades, M. R. and Eades, M. D. (2003). Hyperinsulinemic diseases of civilization: more than just Syndrome X. *Comparative Biochemistry and Physiology Part A*, **136**, 95–112.

Cordain, L., Eaton, S. B., Sebastian, A. *et al.* (2005). Origins and evolution of the Western diet: health implications for the 21st century. *American Journal of Clinical Nutrition*, **81**, 341–54.

Cosmides, L. and Tooby, J. (1997). *Evolutionary Psychology: a Primer*. University of California Santa Barbara: University of California Press.

Costa, D. L. (2000). Understanding the twentieth-century decline in chronic conditions among older men. *Demography*, **37**, 53–72.

Costa, D. L. (2004). The measure of man and older age mortality: evidence from the Gould sample. *Journal of Economic History*, **64**, 1–23.

Craig, P., Halavatau, V., Comino, E. and Caterson, I. (1999). Perception of body size in the Tongan community: differences from and similarities to an Australian sample. *International Journal of Obesity*, **23**, 1288–94.

Critser, G. (2003). *Fat Land: How Americans became the Fattest People in the World*. London: Penguin.

Crowther, R., Dinsdale, H., Rutter, H. and Kyffin, R. (2007). Analysis of the National Childhood Obesity Database 2005–06: South East Public Health Observatory.

Cuijpers, P., Van Straten, A. and Smit, F. (2005). Preventing the incidence of new cases of mental disorders – a meta-analytic review. *Journal of Nervous and Mental Disease*, **193**, 119–25.

Cunningham, A. S. (1995). Breastfeeding: adaptive behavior for child health and longevity. In *Breastfeeding: Biocultural Perspectives*, ed. P. Stuart-Macadam and K. A. Dettwyler. New York: Aldine de Gruyter, pp. 243–64.

Cunningham, G. R. (2006). Testosterone replacement therapy for late-onset hypogonadism. *Nature Clinical Practice Urology*, **3**, 260–7.

Daniel, M., Rowley, K. G., McDermott, R. and O'Dea, K. (2002). Diabetes and impaired glucose tolerance in Aboriginal Australians: prevalence and risk. *Diabetes Research and Clinical Practice*, **57**, 23–33.

Davey Smith, G. and Marmot, M. G. (1991). Trends in mortality in Britain: 1920–1986. *Annals of Nutrition and Metabolism*, **35** (Suppl. 1), 53–63.

Davidson, K., Jonas, B., Dixon, K. and Markovitz, J. (2000). Do depression symptoms predict early hypertension incidence in young adults in the CARDIA study? Coronary artery risk development in young adults. *Archives of Internal Medicine*, **160**, 1495–500.

Davis, A. M., Kreutzer, R., Lipsett, M., King, G. and Shaikh, N. (2006). Asthma prevalence in Hispanic and Asian American ethnic subgroups: Results from the California Healthy Kids Survey. *Pediatrics*, **118**, e363–70.

De Boever, E., De Bacquer, D., Braeckman, L., Baele, L., Rosseneu, M. and De Backer, G. (1995). Relation of fibrinogen to lifestyles and to cardiovascular risk factors in a working population. *International Journal of Epidemiology*, **24**, 915–21.

De Jong, F. H., Oishi, K., Hayes, R. B. *et al.* (1991). Perpipheral hormone levels in controls and patients with prostatic cancer or benign prostatic hyperplasia – results from the Dutch-Japanese case-control study. *Cancer Research*, **51**, 3445–50.

De Laet, C. E. D. H. and Pols, H. A. P. (2000). Fractures in the elderly: epidemiology and demography. *Best Practice and Research in Clinical Endocrinology and Metabolism*, **14**, 171–9.

Delisle, H. F., Rivard, M. and Ekoe, J. M. (1995). Prevalence estimates of diabetes and other cardiovascular risk factors in the two largest Algonquin communities of Quebec. *Diabetes Care*, **18**, 1255–9.

DeMattia, L., Lemont, L. and Meurer, L. (2006). Do interventions to limit sedentary behaviours change behaviour and reduce childhood obesity? A critical review of the literature. *Obesity Reviews*, **8**, 69–81.

Der, G., Batty, D. and Deary, I. J. (2006). Effect of breast feeding on intelligence in children: prospective study, sibling pairs analysis, and meta-analysis. *British Medical Journal*, **333**, 945–50.

Despres, J. P. (2006). Is visceral obesity the cause of the metabolic syndrome? *Annals of Medicine*, **38**, 52–63.

Dettwyler, K. A. (1995a). Beauty and the breast: the cultural context of breastfeeding in the United States. In *Breastfeeding: Biocultural Perspectives*, ed. P. Stuart-Macadam and K. A. Dettwyler. New York: Aldine de Gruyter, pp. 167–215.

Dettwyler, K. A. (1995b). A time to wean: the hominid blueprint for the natural age of weaning in modern human populations. In *Breastfeeding: Biocultural Perspectives*, ed. P. Stuart-Macadam and K. A. Dettwyler. New York: Aldine de Gruyter, pp. 39–74.

Diamond, J. (2003). The double puzzle of diabetes. *Nature*, **423**, 599–602.

Dietz, W. H. (1996). The role of lifestyle in health: the epidemiology and consequences of inactivity. *Proceedings of the Nutrition Society*, **55**, 829–40.

Ding, E. L., Song, Y., Malik, V. and Liu, S. (2006). Sex differences of endogenous sex hormones and risk of type 2 diabetes: a systematic review and meta-analysis. *Journal of the American Medical Association*, **295**, 1288–99.

Dowse, G. K., Zimmet, P. Z. and King, H. (1991). Relationship between prevalence of impaired glucose tolerance and NIDDM in a population. *Diabetes Care*, **14**, 968–74.

Dowse, G. K., Gareeboo, H., Alberti, K. G. M. M. *et al.* (1995). Changes in population cholesterol concentrations and other cardiovascular risk factor levels after five years of the non-communicable disease intervention programme in Mauritius. *British Medical Journal*, **311**, 1255–9.

Drake, A. J. and Walker, B. R. (2004). The intergenerational effects of fetal programming: non-genomic mechanisms for the inheritance of low birth weight and cardiovascular risk. *Journal of Endocrinology*, **180**, 1–16.

Dressler, W. W. (1995). Modeling biocultural interactions: examples from studies of stress and cardiovascular disease. *Yearbook of Physical Anthropology*, **38**, 27–56.

Dressler, W. W., Balieiro, M. C., Ribeiro, R. P. and Dos Santos, J. E. (2005). Cultural consonance and arterial blood pressure in urban Brazil. *Social Science and Medicine*, **61**, 527–40.

Dudley, R. (2000). Evolutionary origins of human alcoholism in primate frugivory. *Quarterly Review of Biology*, **75**, 3–15.

Duggirala, R., Almasy, L., Blangero, J. *et al.* (2003). American Diabetes Association GENNID Study Group: Further evidence for a type 2 diabetes susceptibility locus on chromosome 11q. *Genetic Epidemiology*, **24**, 240–2.

Dunaif, A. (1997). Insulin resistance and the polycystic ovary syndrome: mechanism and implications for pathogenesis. *Endocrine Reviews*, **18**, 774–800.

Dunn, J. E. (1975). Cancer epidemiology in populations of the United States – with an emphasis on Hawaii and California – and Japan. *Cancer Research*, **35**, 3240–5.

Dunne, D. W. and Cooke, A. (2005). A worm's eye view of the immune system: consequences for evolution of human autoimmune disease. *Nature Reviews Immunology*, **5**, 420–6.

Durham, W. H. (1991). *Coevolution: Genes, Culture, and Human Diversity*. Stanford, California: Stanford University Press.

Eaton, S. B. and Eaton, S. B. (1999a). Breast cancer in evolutionary context. In *Evolutionary Medicine*, ed. W. R. Trevathan, E. O. Smith and J. J. McKenna. New York: Oxford University Press, pp. 429–42.

Eaton, S. B. and Eaton, S. B. (1999b). Hunter–gatherers and human health. In *The Cambridge Encyclopedia of Hunters and Gatherers*, ed. R. B. Lee and R. Daly. Cambridge: Cambridge University Press, pp. 449–56.

Eaton, S. B., Konner, M. and Shostak, M. (1988a). Stone agers in the fast lane: chronic degenerative diseases in evolutionary perspective. *American Journal of Medicine*, **84**, 739–49.

Eaton, S. B., Shostak, M. and Konner, M. (1988b). *The Paleolithic Prescription: A Program of Diet and Exercise and a Design for Living*. New York: Harper and Row.

Eaton, S. B., Pike, M. C., Short, R. V. *et al.* (1994). Women's reproductive biology in evolutionary context. *Quarterly Review of Biology*, **69**, 353–67.

Eaton, S. B., Eaton, S. B. and Konner, M. J. (1999). Paleolithic nutrition revisited. In *Evolutionary Medicine*, ed. W. R. Trevathan, E. O. Smith and J. J. McKenna. New York: Oxford University Press, pp. 313–32.

Eaton, S. B., Eaton, S. B. and Cordain, L. (2002). Evolution, diet, and health. In *Human Diet: Its Origin and Evolution*, ed. P. S. Ungar and M. F. Teaford. Westport, Connecticut: Bergin and Garvey, pp. 7–18.

Eckersley, R. (2005). Is modern Western culture a health hazard? *International Journal of Epidemiology*, **35**, 252–8.

Eisenmann, J. C. (2003). Secular trends in variables associated with the metabolic syndrome of North American children and adolescents: review and synthesis. *American Journal of Human Biology*, **15**, 786–94.

Ekbom, A. (2006). The developmental environment and the early origins of cancer. In *Developmental Origins of Health and Disease*, ed. P. Gluckman and M. A. Hanson. Cambridge: Cambridge University Press, pp. 415–25.

Elliott, P., Stamler, J., Nichols, R. *et al.* (1996). Intersalt revisited: further analyses of 24-hour sodium excretion and blood pressure within and across populations. *British Medical Journal*, **312**, 1249–53.

Ellison, G. T. H. and Rees Jones, I. (2002). Social identities and the 'new genetics': scientific and social consequences. *Critical Public Health*, **12**, 265–82.

Ellison, P. T. (1990). Human ovarian function and reproductive ecology: new hypotheses. *American Anthropologist*, **92**, 933–52.

Ellison, P. T. (1996). Developmental influences on adult ovarian hormonal function. *American Journal of Human Biology*, **8**, 725–34.

Ellison, P. T. (1999). Reproductive ecology and reproductive cancers. In *Hormones, Health, and Behavior*, ed. C. Panter-Brick and C. M. Worthman. Cambridge: Cambridge University Press, pp. 184–209.

Ellison, P. T. (2001). *On Fertile Ground: A Natural History of Human Reproduction*. Cambridge, Massachusetts: Harvard University Press.

Ellison, P. T. and Lager, C. (1986). Moderate recreational running is associated with lowered salivary progesterone profiles in women. *American Journal of Obstetrics and Gynecology*, **154**, 1000–3.

Ellison, P. T., Lipson, S. F., O'Rourke, M. T. *et al.* (1993a). Population variation in ovarian function. *The Lancet*, **342**, 433–434.

Ellison, P. T., Panter-Brick, C., Lipson, S. F. and O'Rourke, M. T. (1993b). The ecological context of human ovarian function. *Human Reproduction*, **8**, 2248–58.

Ellison, P. T., Bribiescas, R. G., Bentley, G. R. *et al.* (2002). Population variation in age-related decline in male salivary testosterone. *Human Reproduction*, **17**, 3251–3.

Ellison, P. T. and Jasiénska, G. (2007). Constraint, pathology, and adaptation: how can we tell them apart? *American Journal of Human Biology*, **19**, 622–30.

Elsom, D. M. (1992). *Atmospheric Pollution: A Global Problem*. Oxford: Blackwell.

Emanuel, M. B. (1988). Hay fever, a post industrial revolution epidemic: a history of its growth during the 19th century. *Clinical Allergy*, **18**, 295–304.

Endogenous Hormones and Breast Cancer Collaborative Group (2003). Body mass index, serum sex hormones, and breast cancer risk in postmenopausal women. *Journal of the National Cancer Institute*, **95**, 1218–26.

Erdal, D., Whiten, A., Boehm, C. and Knauft, B. (1994). On human egalitarianism: an evolutionary product of Machiavellian status escalation? *Current Anthropology*, **35**, 175–83.

Ernst, J. and Cacioppo, J. (1999). Lonely hearts: psychological perspectives on loneliness. *Applied and Preventive Psychology*, **8**, 1–22.

Eshed, V., Gopher, A. and Hershkovitz, I. (2006). Tooth wear and dental pathology at the advent of agriculture: new evidence from the Levant. *American Journal of Physical Anthropology*, **130**, 145–59.

Ewald, P. (2002). *Plague Time: The New Germ Theory of Disease*. New York: Anchor Books.

Ewing, R., Schmid, T., Killingsworth, R., Zlot, A. and Raudenbush, S. (2003). Relationship between urban sprawl and physical activity, obesity, and morbidity. *American Journal of Health Promotion*, **5**, 47–57.

Ezzati, M., Vander Hoorn, S., Lawes, C. M. M. *et al.* (2005). Rethinking the "diseases of affluence" paradigm: global patterns of nutritional risk in relation to economic development. *PLoS Medicine*, **2**, e133.

Falkner, B., Sherif, K., Sumner, A. and Kushner, H. (1999). Hyperinsulinism and sex hormones in young adult African Americans. *Metabolism*, **48**, 107–12.

Fall, C. H. D. (2001). Non-industrialised countries and affluence. *British Medical Bulletin*, **60**, 33–50.

Fanaro, S., Chierici, R., Guerrini, P. and Vigi, V. (2003). Intestinal microflora in early infancy: composition and development. *Acta Paeditrica*, **441** (Suppl.), 48–55.

FAO/WHO/UNU Expert Consultation (1985). *Energy and Protein Requirements*. Geneva: World Health Organization.

Fee, M. (2006). Racializing narratives: obesity, diabetes and the "Aboriginal" thrifty genotype. *Social Science and Medicine*, **62**, 2988–97.

Fein, S. B. and Roe, B. (1998). The effect of work status on initiation and duration of breast-feeding. *American Journal of Public Health*, **88**, 1042–6.

Feldman, H. A., Goldstein, I., Hatzichkristou, D. G., Krane, R. J. and McKinlay, J. B. (1994). Impotence and its medical and psychosocial correlates: results of the Massachusetts Male Aging Study. *Journal of Urology*, **151**, 54–61.

Fiennes, R. (1978). *Zoonoses and the Origins and Ecology of Human Disease*. London: Academic Press.

Fildes, V. (1995). The culture and biology of breastfeeding: An historical review of western Europe. In *Breastfeeding: Biocultural Perspectives*, ed. P. Stuart-Macadam and K. A. Dettwyler. New York: Aldine de Gruyter, pp. 101–26.

Finch, C. E. and Crimmins, E. M. (2004). Inflammatory exposure and historical changes in human life-spans. *Science*, **305**, 1736–9.

Flaherman, V. and Rutherford, G. W. (2006). A meta-analysis of the effect of high weight on asthma. *Archives of Disease in Childhood*, **91**, 334–9.

Flegal, K. M., Carroll, M. D., Kuczmarski, R. J. and Johnson, C. (1998). Overweight and obesity in the United States: prevalence and trends, 1960–1994. *International Journal of Obesity*, **22**, 39–47.

Fleming, J. O. and Cook, T. D. (2006). Multiple sclerosis and the hygiene hypothesis. *Neurology*, **67**, 2085–6.

Fletcher, E. S., Rugg-Gunn, A. J., Matthews, J. N. S. *et al.* (2004). Changes over 20 years in macronutrient intake and body mass index in 11- to 12-year-old adolescents living in Northumberland. *British Journal of Nutrition*, **92**, 321–33.

Fogel, R. W. and Costa, D. L. (1997). A theory of technophysio evolution, with some implications for forecasting population, health care costs, and pension costs. *Demography*, **34**, 49–66.

Foley, R. A. (1993). The influence of seasonality on hominid evolution. In *Seasonality and Human Ecology*, ed. S. J. Ulijaszek and S. S. Strickland. Cambridge: Cambridge University Press, pp. 17–37.

Foley, R. A. (1996). The adaptive legacy of human evolution: A search for the environment of evolutionary adaptedness. *Evolutionary Anthropology*, **4**, 194–203.

Foley, R. A. and Lee, P. C. (1991). Ecology and energetics of encephalisation in hominid evolution. *Philosophical Transactions of the Royal Society of London B*, **334**, 223–32.

Ford, E. S., Giles, W. H. and Dietz, W. H. (2002). Prevalence of the metabolic syndrome among US adults: findings from the Third National Health and Nutrition Examination Survey. *Journal of the American Medical Association*, **287**, 356–9.

Forouhi, N. and Sattar, N. (2006). CVD risk factors and ethnicity – a homogeneous relationship? *Atherosclerosis Supplements*, **7**, 11–19.

Fox, C. S., Pencina, M. J., Meigs, J. B., Vasan, R. S., Levitzky, Y. S. and D'Agostino, R. B. (2006). Trends in the incidence of type 2 diabetes mellitus from the 1970s to the 1990s. *Circulation*, **113**, 2914–18.

Friedman, J. M. (2004). Modern science versus the stigma of obesity. *Nature Medicine*, **10**, 563–9.

Friedman, N. J. and Zeiger, R. S. (2005). The role of breast-feeding in the development of allergies and asthma. *Journal of Allergy and Clinical Immunology*, **115**, 1238–48.

Frisch, R. E., Wyshak, G., Albright, N. L. *et al.* (1985). Lower prevalence of breast cancer and cancers of the reproductive system among former college athletes compared to non-athletes. *British Journal of Cancer*, **52**, 885–91.

Frost, G. and Dornhorst, A. (2001). Starting the day the right way. *The Lancet*, **357**, 736–7.

Fullerton, S. M., Bartoszewicz, A., Ybazeta, G. *et al.* (2002). Geographic and haplotype structure of putative type 2 diabetes susceptibility variants at the Calpain-10 locus. *American Journal of Human Genetics*, **70**, 1096–106.

Furberg, A. S., Jasieńska, G., Bjurstam, N. *et al.* (2005). Metabolic and hormonal profiles: HDL cholesterol as a plausible biomarker of breast cancer risk. The Norwegian EBBA study. *Cancer Epidemiology, Biomarkers and Prevention*, **14**, 33–40.

Fuster, V., Badimon, L., Badimon, J. J. and Chesebro, J. H. (1992a). The pathogenesis of coronary artery disease and the acute coronary syndromes: Part I. *New England Journal of Medicine*, **326**, 242–50.

Fuster, V., Badimon, L., Badimon, J. J. and Chesebro, J. H. (1992b). The pathogenesis of coronary artery disease and the acute coronary syndromes: Part II. *New England Journal of Medicine*, **326**, 310–18.

Gage, T. B. (2005). Are modern environments really bad for us?: revisiting the demographic and epidemiologic transitions. *Yearbook of Physical Anthropology*, **48**, 96–117.

Gale, E. A. M. (2002). A missing link in the hygiene hypothesis? *Diabetologia*, **45**, 588–94.

Gallou-Kabani, C. and Junien, C. (2005). Nutritional epigenomics of metabolic syndrome. *Diabetes*, **54**, 1899–906.

Galloway, A. (1997). The cost of reproduction and the evolution of postmenopausal osteoporosis. In *The Evolving Female: A Life-History Perspective*, ed. M. E. Morbeck, A. Galloway and A. L. Zihlman. Princeton: Princeton University Press, pp. 132–46.

Gandy, M. and Zumla, A. (2002). The resurgence of disease: social and historical perspectives on the 'new' tuberculosis. *Social Science and Medicine*, **55**, 385–96.

Gann, P. H., Hennekens, C. H., Ma, J., Longcope, C. and Stampfer, M. J. (1996). Prospective study of sex hormone levels and risk of prostate cancer. *Journal of the National Cancer Institute*, **88**, 1118–26.

Gapstur, S. M., Gann, P. H., Kopp, P., Colangelo, L., Longcope, C. and Liu, K. (2002). Serem androgen concentrations in young men: a longitudinal analysis of associations with age, obesity, and race. The CARDIA Male Hormone study. *Cancer Epidemiology, Biomarkers and Prevention*, **11**, 1041–7.

Gardell, B. (1987). Efficiency and health hazards in mechanized work. In *Work Stress*, ed. J. C. Quick, R. Bhagat, J. Dalton and J. D. Quick. New York: Praeger, pp. 50–71.

Gaulin, S. J. C. (1980). Sexual dimorphism in the human post-reproductive life-span: possible causes. *Journal of Human Evolution*, **9**, 227–32.

Gibson, M. A. and Mace, R. (2005). Helpful grandmothers in rural Ethiopia: A study of the effect of kin on child survival and growth. *Evolution and Human Behavior*, **26**, 469–82.

Glassman, A. H., Helzer, J. E., Covey, L. S., Stetner, F., Tipp, J. E. and Johnson, J. (1990). Smoking, smoking cessation, and major depression. *Journal of the American Medical Association*, **264**, 1546–9.

Gleibermann, L. (1973). Blood pressure and dietary salt in human populations. *Ecology of Food and Nutrition*, **2**, 143–56.

Gluckman, P. and Hanson, M. (2005). *The Fetal Matrix: Evolution, Development and Disease*. Cambridge: Cambridge University Press.

Goldin, B. R., Adlercreutz, H., Gorbach, S. L. *et al.* (1986). The relationship between estrogen levels and diets of Caucasian American and Oriental immigrant women. *American Journal of Clinical Nutrition*, **44**, 945–53.

Gould, D. C., Petty, R. and Jacobs, H. S. (2000). The male menopause - does it exist? *British Medical Journal*, **320**, 858–61.

Gracey, M. and Spargo, R. M. (1987). The state of health of Aborigines in the Kimberley region. *Medical Journal of Australia*, **146**, 200–4.

Gray, P. B., Kruger, A., Huisman, H. M., Wissing, M. P. and Vorster, H. H. (2006). Predictors of South African male testosterone levels: the THUSA study. *American Journal of Human Biology*, **18**, 123–32.

Gray-Donald, K., Jacobs-Starkey, L. and Johnson-Down, L. (2000). Food habits of Canadians: reduction in fat intake over a generation. *Canadian Journal of Public Health*, **91**, 381–5.

Greenwood, D. C., Muir, K. R., Packham, C. J. and Madeley, R. J. (1996). Coronary heart disease: a review of the role of psychosocial stress and social support. *Journal of Public Health Medicine*, **18**, 221–31.

Griffiths, K., Denis, L., Turkes, A. and Morton, M. S. (1998). Phytoestrogens and diseases of the prostate gland. *Baillière's Clinical Endocrinology and Metabolism*, **12**, 625–47.

Gröland, M. M., Lehtonen, O.-P., Eerola, E. and Kero, P. (1999). Fecal microflora in healthy infants born by different methods of delivery: permanent changes in intestinal flora after Cesarian delivery. *Journal of Pediatric Gastroenterology and Nutrition*, **28**, 19–25.

Gross, L. S., Ford, E. S. and Liu, S. (2004). Increased consumption of refined carbohydrates and the epidemic of type 2 diabetes in the United States: an ecologic assessment. *American Journal of Clinical Nutrition*, **79**, 774–9.

Grossman, H., Bergmann, C. and Parker, S. (2006). Dementia: a brief review. *Mount Sinai Journal of Medicine*, **73**, 985–92.

Grundy, J., Matthews, S., Bateman, B., Dean, T. and Arshad, S. H. (2002). Rising prevalence of allergy to peanuts in children: data from two sequential cohorts. *Journal of Allergy and Clinical Immunology*, **110**, 784–9.

Gu, K., Cowie, C. C. and Harris, M. I. (1998). Mortality in adults with and without diabetes in a national cohort of the US population, 1971–1993. *Diabetes Care*, **21**, 1138–45.

Guarner, F., Bourdet-Sicard, R., Brandtzaeg, P. *et al.* (2006). Mechanisms of disease: the hygiene hypothesis. *Nature Clinical Practice Gastroenterology and Hepatology*, **3**, 275–84.

Guest, C. S., O'Dea, K., Hopper, J. L. and Larkins, R. G. (1993). Hyperinsulinaemia and obesity in Aborigines of south-eastern Australia, with comparisons from rural and urban Europid populations. *Diabetes Research and Clinical Practice*, **20**, 155–64.

Guise, J. M., Austin, D. and Morris, C. D. (2005). Review of case-control studies related to breastfeeding and reduced risk of childhood leukemia. *Pediatrics*, **116**, e724–31.

Guyton, A. C. (1986). *Textbook of Medical Physiology*. Philadelphia: WB Saunders.

Haffner, S. M. (2003). Insulin resistance, inflammation, and the prediabetic state. *American Journal of Cardiology*, **92** (Suppl.), 18J–26J.

Hagen, E. H. (1999). The functions of postpartum depression. *Evolution and Human Behavior*, **20**, 325–59.

Hajat, S., Haines, A. P., Atkinson, R. W., Bremner, S., Anderson, H. R. and Emberlin, J. (2001). Association between air pollution and daily consultations with General Practitioners for allergic rhinitis in London, United Kingdom. *American Journal of Epidemiology*, **153**, 704–14.

Hales, C. N. and Barker, D. J. P. (1992). Type 2 (non-insulin-dependent) diabetes mellitus: the thrifty phenotype hypothesis. *Diabetologia*, **35**, 595–601.

Hales, C.N. and Barker, D.J.P. (2001). The thrifty phenotype hypothesis. *British Medical Bulletin*, **60**, 5–20.

Hall, W.D., Clark, L.T., Wenger, N.H. *et al.* (2003). The metabolic syndrome in African Americans: a review. *Ethnicity and Disease*, **13**, 414–28.

Hamilton, G. (2005). Filthy friends. *New Scientist*, **2495**, 35–9.

Hankinson, S.E., Willett, W.C., Manson, J.E. *et al.* (1995). Alcohol, height, and adiposity in relation to estrogen and prolactin levels in postmenopausal women. *Journal of the National Cancer Institute*, **87**, 1297–302.

Hannaford, P.C., Selvaraj, S., Elliott, A.M., Angus, V., Iversen, L. and Lee, A.J. (2007). Cancer risk among users of oral contraceptives: cohort data from the Royal College of General Practioners' oral contraception study. *British Medical Journal*, **335**, 651–4A.

Hanson, R.L., Ehm, M.G., Pettitt, D.J. *et al.* (1998). An autosomal genomic scan for loci linked to type II diabetes mellitus and body-mass index in Pima Indians. *American Journal of Human Genetics*, **63**, 1130–8.

Harder, T., Bergmann, R., Kallischnigg, G. and Plagemann, A. (2005). Duration of breastfeeding and risk of overweight: A meta-analysis. *American Journal of Epidemiology*, **162**, 397–403.

Harder, T., Rodekamp, E., Schellong, K., Duderhausen, J.W., Plagemann, A. (2007). Birth weight and subsequent risk of type 2 diabetes: a meta-analysis. *American Journal of Epidemiology*, **165**, 849–57.

Hardy, R. and Kuh, D. (2002). Change in psychological and vasomotor symptom reporting during the menopause. *Social Science and Medicine*, **55**, 1975–88.

Harman, S.M., Metter, E.J., Tobin, J.D., Pearson, J. and Blackman, M.R. (2001). Longitudinal effects of aging on serum total and free testosterone levels in healthy men. *Journal of Clinical Endocrinology and Metabolism*, **86**, 724–31.

Harris, D.R. and Hillman, G.C., eds. (1989). *Foraging and Farming: The Evolution of Plant Exploitation*. London: Unwin Hyman.

Harris, M.I., Goldstein, D.E., Flegal, K.M. *et al.* (1998). Prevalence of diabetes, impaired fasting glucose, and impaired glucose tolerance in US adults. *Diabetes Care*, **21**, 518–24.

Harris, R., Tobias, M., Jeffreys, M., Waldegrave, K., Karlsen, S. and Nazroo, J. (2006). Effects of self-reported racial discrimination and deprivation on Māori health and inequalities in New Zealand: cross-sectional study. *The Lancet*, **367**, 2005–9.

Harris, S.B., Gittelsohn, J., Hanley, A.J.G. *et al.* (1997). The prevalence of NIDDM and associated risk factors in Native Canadians. *Diabetes Care*, **20**, 185–7.

Harrison, G.A. (1973). The effects of modern living. *Journal of Biosocial Science*, **5**, 217–28.

Harrison, G.A., Tanner, J.A., Pilbeam, D.R. and Baker, P.T. (1988). *Human Biology: An Introduction to Human Evolution, Variation, Growth, and Adaptability*. Oxford: Oxford University Press.

Hart, R., Hickey, M. and Franks, S. (2004). Definitions, prevalence and symptoms of polycystic ovaries and polycystic ovary syndrome. *Best Practice and Research in Clinical Obstetrics and Gynaecology*, **18**, 671–683.

Harvie, M., Hooper, L. and Howell, A. H. (2003). Central obesity and breast cancer risk: a systematic review. *Obesity Reviews*, **4**, 157–73.

Hawkes, K. (2003). Grandmothers and the evolution of human longevity. *American Journal of Human Biology*, **15**, 380–400.

He, F. J., Nowson, C. A. and MacGregor, G. A. (2006). Fruit and vegetable consumption and stroke: meta-analysis of cohort studies. *The Lancet*, **367**, 320–6.

Hegele, R. A. and Bartlett, L. C. (2003). Genetics, environment and type 2 diabetes in the Oji-Cree population of northern Ontario. *Canadian Journal of Diabetes*, **27**, 256–61.

Hegele, R. A., Cao, H., Hanley, A. J. G., Zinman, B., Harris, S. B. and Anderson, C. M. (2000). Clinical utility of HNF1A genotyping for diabetes in Aboriginal Canadians. *Diabetes Care*, **23**, 775–8.

Heim, C., Ehlert, U. and Hellhammer, D. H. (2000). The potential role of hypocortisolism in the pathophysiology of stress-related bodily disorders. *Psychoneuroendocrinology*, **25**, 1–35.

Heinig, M. J. (2001). Host defense benefits of breastfeeding for the infant. *Pediatric Clinics of North America*, **48**, 105–23.

Heinig, M. J. and Dewey, K. G. (1997). Health effects of breast feeding for mothers: a critical review. *Nutrition Research Reviews*, **10**, 35–56.

Helman, C. G. (1994). *Culture, Health and Illness*. Oxford: Butterworth-Heinemann.

Helmchen, L. A. and Henderson, R. M. (2004). Changes in the distribution of body mass index of white US men, 1890–2000. *Annals of Human Biology*, **31**, 174–81.

Hemingway, H. and Marmot, M. (1999). Psychosocial factors in the aetiology and prognosis of coronary heart disease: systematic review of prospective cohort studies. *British Medical Journal*, **318**, 1460–7.

Henderson, B. E., Ross, R. K., Pike, M. C. and Cassagrande, J. T. (1982). Endogenous hormones as a major factor in human cancer. *Cancer Research*, **42**, 3232–9.

Hendrix, S. L., Wassertheil-Smoller, S., Johnson, K. C. *et al.* for the WHI Investigators (2006). Effects of conjugated equine estrogen on stroke in the Women's Health Initiative. *Circulation*, **113**, 2425–34.

Henry, C. J. K., Lightowler, H. J. and Al-Hourani, H. M. (2004). Physical activity and levels of inactivity in adolescent females aged 11–16 years in the United Arab Emirates. *American Journal of Human Biology*, **16**, 346–53.

Herbst, K. L. and Bhasin, S. (2004). Testosterone action on skeletal muscle: anabolic and catabolic signals. *Current Opinion in Clinical Nutrition and Metabolic Care*, **7**, 271–7.

Heuveline, P., Guillot, M. and Gwatkin, D. R. (2002). The uneven tides of the health transition. *Social Science and Medicine*, **55**, 313–22.

Hill, K. and Hurtado, A. M. (1991). The evolution of premature reproductive senescence and menopause in human females: an evaluation of the 'grandmother hypothesis'. *Human Nature*, **2**, 313–50.

Hill, K. and Hurtado, A. M. (1996). *Ache Life History*. New York: Aldine de Gruyter.

Hill, K. and Kaplan, H. (1999). Life history traits in humans: theory and empirical studies. *Annual Review of Anthropology*, **28**, 397–430.

Hoffjan, S., Nicolae, D. and Ober, C. (2003). Association studies for asthma and atopic diseases: a comprehensive review of the literature. *Respiratory Research*, **4**, 14.

Holland, T. D. and O'Brien, M. J. (1997). Parasites, porotic hyperostosis, and the implications of changing perspectives. *American Antiquity*, **62**, 183–93.

Holman, R. C., Curns, A. T., Kaufman, S. F., Cheek, J. E., Pinner, R. W. and Schonberger, L. B. (2001). Trends in infectious disease hospitalizations among American Indians and Alaska Natives. *American Journal of Public Health*, **91**, 425–31.

Horikawa, Y., Oda, N., Cox, N. J. *et al.* (2000). Genetic variation in the gene encoding calpain-10 is associated with type 2 diabetes mellitus. *Nature Genetics*, **26**, 163–75.

Howard, B. V., Lee, E. T., Cowan, L. D. *et al.* (1999). Rising tide of cardiovascular disease in American Indians. *Circulation*, **99**, 2389–95.

Hrdy, S. B. (2000). *Mother Nature*. London: Vintage.

Hsing, A. W. (1996). Hormones and prostate cancer: where do we go from here? *Journal of the National Cancer Institute*, **88**, 1093–5.

Hsing, A. W. and Devesa, S. S. (2001). Trends and patterns of prostate cancer: What do they suggest? *Epidemologic Reviews*, **23**, 3–13.

Hsing, A. W., Gao, Y. T., Chua, S., Deng, J. and Stanczyk, F. Z. (2003). Insulin resistance and prostate cancer risk. *Journal of the National Cancer Institute*, **95**, 67–71.

Hudson, J. I., Hiripi, E., Pope, H. G. and Kessler, R. C. (2007). The prevalence and correlates of eating disorders in the National Comorbidity Survey Replication. *Biological Psychiatry*, **61**, 348–58.

Hulley, S., Grady, D., Bush, T. *et al.* (1998). Randomized trial of estrogen plus progestin for secondary prevention of coronary heart disease in postmenopausal women. *Journal of the American Medical Association*, **280**, 605–13.

Hurtado, A. M., Arenas, I. and Hill, K. (1996). Evolutionary contexts of chronic allergic disease: the Hiwi of Venezuala. *Human Nature*, **8**, 1–20.

Hussain, A., Rahim, M. A., Azad Khan, A. K., Ali, S. M. K. and Vaaler, S. (2005). Type 2 diabetes in rural and urban populations: diverse prevalence and associated risk factors in Bangladesh. *Diabetic Medicine*, **22**, 931–6.

Huxley, R., Neil, A. and Collins, R. (2002). Unravelling the fetal origins hypothesis: is there really an inverse association between birthweight and subsequent blood pressure? *The Lancet*, **360**, 659–65.

Huxley, R., Owens, J. F., Whincup, P. H., Cook, D. G., Colman, S. and Collins, R. (2004). Birth weight and subsequent cholesterol levels: exploration of the "fetal origins" hypothesis. *Journal of the American Medical Association*, **292**, 2755–64.

Huyghe, E., Matsuda, T. and Thonneau, P. (2003). Increasing incidence of testicular cancer worldwide: a review. *Journal of Urology*, **170**, 5–11.

Iglowstein, I., Jenni, O. G., Molinari, L. and Largo, R. H. (2003). Sleep duration from infancy to adolescence: reference values and generational trends. *Pediatrics*, **111**, 302–7.

Irons, W. (1998). Adaptively relevant environments versus the environment of evolutionary adaptedness. *Evolutionary Anthropology*, **6**, 198–204.

ISAAC Steering Committee (1998). Worldwide variation in the prevalence of symptoms of asthma, allergic rhinoconjunctivitis, and atopic eczema: ISAAC. *The Lancet*, **351**, 1225–32.

Jackson, M. A., Kovi, J., Heshmat, M. Y. *et al.* (1980). Characterization of prostatic carcinoma among blacks: A comparison between a low-incidence area, Ibadan, Nigeria, and a high-incidence area, Washington, DC. *Prostate*, **1**, 185–205.

James, G. D., Jenner, D. A., Harrison, G. A. and Baker, P. T. (1985). Differences in catecholamine excretion rates, blood pressure and lifestyle among young Western Samoan men. *Human Biology*, **57**, 635–47.

James, G. D., Crews, D. E. and Pearson, J. (1989). Catecholamines and stress. In *Human Population Biology*, ed. M. A. Little and J. D. Haas. Oxford: Oxford University Press.

James, G. D., Broege, P. A. and Schlussel, Y. R. (1996). Assessing cardiovascular risk and stress-related blood pressure variability in young women employed in waged jobs. *American Journal of Human Biology*, **8**, 743–9.

James, W. P. T. (2002). Will feeding mothers prevent the Asian metabolic syndrome epidemic? *Asia Pacific Journal of Clinical Nutrition*, **11** (Suppl.), S516–23.

James, W. P. T. (2005). Assessing obesity: are ethnic differences in body mass index and waist classification criteria justified? *Obesity Reviews*, **6**, 179–81.

Janes, C. R. (1990). Migration, changing gender roles and stress: the Samoan case. *Medical Anthropology*, **12**, 145–67.

Jasieńska, G. and Ellison, P. T. (1998). Physical work causes suppression of ovarian function in women. *Proceedings of the Royal Society of London B*, **265**, 1847–51.

Jasieńska, G. and Ellison, P. T. (2004). Energetic factors and seasonal changes in ovarian function in women from rural Poland. *American Journal of Human Biology*, **16**, 563–80.

Jasieńska, G. and Thune, I. (2001a). Lifestyle, hormones, and risk of breast cancer. *British Medical Journal*, **322**, 586–7.

Jasieńska, G. and Thune, I. (2001b). Lifestyle, progesterone, and risk of breast cancer – reply. *British Medical Journal*, **323**, 1002.

Jasieńska, G., Thune, I. and Ellison, P. T. (2000). Energetic factors, ovarian steroids and the risk of breast cancer. *European Journal of Cancer Prevention*, **9**, 231–9.

Jasieńska, G., Thune, I. and Ellison, P. T. (2006a). Fatness at birth predicts adult susceptibility to ovarian suppression: an empirical test of the Predictive Adaptive Response hypothesis. *Proceedings of the National Academy of Science*, **103**, 12759–62.

Jasieńska, G., Ziomkiewicz, A., Thune, I., Lipson, S. F. and Ellison, P. T. (2006b). Habitual physical activity and estradiol levels in women of reproductive age. *European Journal of Cancer Prevention*, **15**, 439–45.

Jenike, M. R. (2001). Nutritional ecology: diet, physical activity and body size. In *Hunter–Gatherers: An Interdisciplinary Perspective*, ed. C. Panter-Brick, R. H. Layton and P. Rowley-Conwy. Cambridge: Cambridge University Press, pp. 205–39.

Jenner, D. A., Reynolds, V. and Harrison, G. A. (1980). Catecholamine excretion rates and occupation. *Ergonomics*, **23**, 237–46.

Jenner, D. A., Harrison, G. A. and Prior, I. A. M. (1987). Catecholamine excretion in Tokelauans living in three different environments. *Human Biology*, **59**, 165–72.

Jin, B., Turner, L., Zhou, Z., Zhou, E. L. and Handelsman, D. J. (1999). Ethnicity and migration as determinants of human prostate size. *Journal of Clinical Endocrinology and Metabolism*, **84**, 3613–19.

Jonas, B. S., Franks, P. and Ingram, D. D. (1997). Are symptoms of anxiety and depression risk factors for hypertension? Longitudinal evidence from the National Health and Nutrition Examination Survey I: epidemiological follow-up study. *Archives of Family Medicine*, **6**, 43–9.

Julian, D. G. and Cowan, J. C. (1992). *Cardiology*. London: Baillière Tindall.

Kaaks, R., Lukanova, A. and Kurzer, M. S. (2002). Obesity, endogenous hormones, and endometrial cancer risk: a synthetic review. *Cancer Epidemiology, Biomarkers and Prevention*, **11**, 1531–43.

Kajantie, E. and Phillips, D. I. W. (2006). The effects of sex and hormonal status on the physiological response to acute psychosocial stress. *Psychoneuroendocrinology*, **31**, 151–78.

Kaliora, A. C., Dedoussis, G. V. Z. and Schmidt, H. (2006). Dietary antioxidants in preventing atherogenesis. *Atherosclerosis*, **187**, 1–17.

Kalliomäki, M., Salminen, S., Arvilommi, H., Kero, P., Koskinen, P. and Isolauri, E. (2001). Probiotics in primary prevention of atopic disease: a randomised placebo-controlled trial. *The Lancet*, **357**, 1076–9.

Kannel, W. B. (1987). New perspectives on cardiovascular risk factors. *American Heart Journal*, **114**, 213–19.

Kant, A. K. and Graubard, B. I. (2004). Eating out in America, 1987–2000: trends and nutritional correlates. *Preventive Medicine*, **38**, 243–9.

Karasek, R. A. (1979). Job demands, job decision latitude, and mental strain: implications for job redesign. *Administrative Science Quarterly*, **24**, 285–308.

Kaufman, J. S. and Hall, S. A. (2003). The slavery hypertension hypothesis: dissemination and appeal of a modern race theory. *Epidemiology*, **14**, 111–18.

Kaur, H., Choi, W. S., Mayo, M. S. and Harris, K. J. (2003). Duration of television watching is associated with increased body mass index. *Journal of Pediatrics*, **143**, 506–11.

Keighley, E. D., McGarvey, S. T., Quested, C., McCuddin, C., Viali, S. and Maga, U. A. (2007). Nutrition and health in modernizing Samoans: temporal trends and adaptive perspectives. In *Health Changes in the Asia-Pacific Region: Biocultural and Epidemiological Approaches*, ed. R. Ohtsuka and S. J. Ulijaszek. Cambridge: Cambridge University Press, pp. 147–91.

Keil, J. E., Sutherland, S. E., Knapp, R. G. and Tyroler, H. A. (1992). Does equal socioeconomic status in black and white men mean equal risk of mortality? *American Journal of Public Health*, **82**, 1133–6.

Kemp, A. and Kakakios, A. (2004). Asthma prevention: breast is best? *Journal of Paediatrics and Child Health*, **40**, 337–9.

Kennedy, G. E. (2005). From the ape's dilemma to the weanling's dilemma: early weaning and its evolutionary context. *Journal of Human Evolution*, **48**, 123–45.

Kent, S. (1986). The influence of sedentism and aggregation on porotic hyperostosis and anemia: a case study. *Man*, **21**, 605–36.

Kero, J., Gissler, M., Hemminki, E. and Isolauri, E. (2001). Could T(h)1 and T(h)2 diseases coexist? Evaluation of asthma incidence in children with coeliac disease, type 1 diabetes, or rheumatoid arthritis: a register study. *Journal of Allergy and Clinical Immunology*, **108**, 781–3.

Key, T. J. A., Chen, J., Wang, D. Y., Pike, M. C. and Boreham, J. (1990). Sex hormones in women in rural China and in Britain. *British Journal of Cancer*, **62**, 631–6.

Kim, S. Y., Moon, S. and Popkin, B. M. (2000). The nutrition transition in South Korea. *American Journal of Clinical Nutrition*, **71**, 44–53.

King, H., Aubert, R. E. and Herman, W. H. (1998). Global burden of diabetes, 1995–2025. *Diabetes Care*, **21**, 1414–31.

Kleinman, A. and Good, B. (1985). Introduction: culture and depression. In *Culture and Depression: Studies in the Anthropology and Cross-Cultural Psychiatry of Affect and Disorder*, ed. A. Kleinman and B. Good. Berkeley: University of California Press, pp. 1–33.

Kleinman, A. M. (1977). Depression, somatization and the "new cross-cultural" psychiatry. *Social Science and Medicine*, **11**, 3–10.

Klerman, G. L. and Weissman, M. M. (1989). Increasing rates of depression. *Journal of the American Medical Association*, **261**, 2229–35.

Kliks, M. M. (1983). Paleoparasitology: on the origins and impact of human-helminth relationships. In *Human Ecology and Infectious Disease*, ed. N. A. Croll and J. H. Cross. New York: Academic Press.

Knowler, W. C. (1978). Diabetes incidence and prevalence in Pima Indians: a 19-fold greater incidence than in Rochester, Minnesota. *American Journal of Epidemiology*, **108**, 497–505.

Knowler, W. C., Pettit, D. J., Savage, P. J., *et al.* (1983). Diabetes mellitus in the Pima Indians: genetic and evolutionary considerations. *American Journal of Physical Anthropology*, **62**, 107–14.

Kohen-Avramoglu, R., Theriault, A. and Adeli, K. (2003). Emergence of the metabolic syndrome in childhood: an epidemiological overview and mechanistic link to dyslipidemia. *Clinical Biochemistry*, **36**, 413–20.

Kohler, H. P., Billari, F. C. and Ortega, J. A. (2002). The emergence of lowest-low fertility in Europe during the 1990s. *Population and Development Review*, **28**, 641–86.

Konner, M. and Worthman, C. M. (1980). Nursing frequency, gonadal function, and birth spacing among the !Kung hunter–gatherers. *Science*, **207**, 788–91.

Kraft, P., Pharoah, P., Chanock, S. J. *et al.* (2005). Genetic variation in the HSD17B1 gene and risk of prostate cancer. *PLoS Genetics*, **1**, e68.

Kramer, M. S. (1987). Determinants of low birth weight: methodological assessment and meta-analysis. *Bulletin of the World Health Organization*, **65**, 663–737.

Kramer, M. S. and Kakuma, R. (2002). *The Optimal Duration of Exclusive Breastfeeding: A Systematic Review*. Geneva: World Health Organization.

Krämer, U., Heinrich, J., Wjst, M. and Wichmann, H.-E. (1999). Age of entry to day nursery and allergy in later childhood. *The Lancet*, **352**, 450–4.

Krantz, D. S., Contrada, R. J., Hill, R. and Friedler, E. (1988). Environmental stress and biobehavioral antecedents of coronary heart disease. *Journal of Consulting and Clinical Psychology*, **56**, 333–41.

Krause, I.-B. (1989). Sinking heart: a Punjabi communication of distress. *Social Science and Medicine*, **29**, 563–75.

Kubzansky, L. D. and Kawachi, I. (2000). Going to the heart of the matter: do negative emotions cause coronary heart disease? *Journal of Psychosomatic Research*, **48**, 323–37.

Kuh, D. and Hardy, R., eds. (2002). *A Life Course Approach to Women's Health*. Oxford: Oxford University Press.

Kuh, D., Ben-Shlomo, Y., Lynch, J., Hallqvist, J. and Power, C. (2003). Life course epidemiology. *Journal of Epidemiology and Community Health*, **57**, 778–83.

Kukkonen, K., Savilahti, E., Haahtela, T. *et al.* (2007). Probiotics and prebiotic galacto-oligosaccharides in the prevention of allergic diseases: a randomized, double-blind, placebo-controlled trial. *Journal of Allergy and Clinical Immunology*, **119**, 192–8.

Kuller, L. H., Meilahn, E. N., Cauley, J. A., Gutai, J. P. and Matthews, K. A. (1994). Epidemiologic studies of menopause: changes in risk factors and disease. *Experimental Gerontology*, **29**, 495–509.

Künzli, N., McConnel, R., Bates, D. *et al.* (2003). Breathless in Los Angeles: the exhausting search for clean air. *American Journal of Public Health*, **93**, 1494–99.

Kuzawa, C. W. (1998). Adipose tissue in human infancy and childhood: an evolutionary perspective. *Yearbook of Physical Anthropology*, **41**, 177–210.

Kuzawa, C. W. (2005). Fetal origins of developmental plasticity: are fetal cues reliable predictors of future nutritional environments? *American Journal of Human Biology*, **17**, 5–21.

Lager, C. and Ellison, P. T. (1990). Effect of moderate weight loss on ovarian function assessed by salivary progesterone measurements. *American Journal of Human Biology*, **2**, 303–12.

Lake, A. and Townshend, T. (2006). Obesogenic environments: exploring the built and food environments. *Journal of the Royal Society for the Promotion of Health*, **126**, 262–7.

LaMonte, M. J., Barlow, C. E., Jurca, R., Kampert, J. B., Church, T. S. and Blair, S. N. (2005). Cardiorespiratory fitness is inversely associated with the incidence of metabolic syndrome. A prospective study of men and women. *Circulation*, **112**, 505–12.

Lancet Editorial (2006). A plea to abandon asthma as a disease concept. *The Lancet*, **368**, 705.

Lane, D., Beevers, D. G. and Lip, G. Y. H. (2002). Ethnic differences in blood pressure and the prevalence of hypertension in England. *Journal of Human Hypertension*, **16**, 267–73.

Larsen, C. S. (1995). Biological changes in human populations with agriculture. *Annual Review of Anthropology*, **24**, 185–213.

Larsen, C. S. (2002). Post-Pleistocene human evolution: bioarcheology of the agricultural transition. In *Human Diet: Its Origin and Evolution*, ed. P. S. Ungar and M. F. Teaford. Westport, Connecticut: Bergin and Garvey, pp. 19–36.

Larsen, C. S. (2003). Animal source foods and human health during evolution. *Journal of Nutrition*, **133**, 3893S–7S.

Lasker, G. (1969). Human biological adaptability: the ecological approach in physical anthropology. *Science*, **166**, 1480–6.

La Vecchia, C. and Bosetti, C. (2004). Benefits and risks of oral contraceptives on cancer. *European Journal of Cancer Prevention*, **13**, 467–70.

Lawlor, D. A., Davey Smith, G., Leon, D. A., Sterne, J. A. C. and Ebrahim, S. (2002). Secular trends in mortality by stroke subtype in the 20th century: a retrospective analysis. *The Lancet*, **360**, 1818–23.

Le Souëf, P. N., Goldblatt, J. and Lynch, N. R. (2000). Evolutionary adaptation of inflammatory immune responses in human beings. *The Lancet*, **356**, 242–4.

Le Souëf, P. N., Candelaria, P. and Goldblatt, J. (2006). Evolution and respiratory genetics. *European Respiratory Journal*, **28**, 1258–63.

Lee, R. B. (1979). *The !Kung San: Men, Women and Work in a Foraging Society*. Cambridge: Cambridge University Press.

Leidy, L. E. (1999). Menopause in evolutionary perspective. In *Evolutionary Medicine*, ed. W. R. Trevathan, E. O. Smith and J. J. McKenna. New York: Oxford University Press, pp. 407–27.

Leidy Sievert, L. (2001). Aging and reproductive senescence. In *Reproductive Ecology and Human Evolution*, ed. P. T. Ellison. New York: Aldine de Gruyter, pp. 267–92.

Leidy Sievert, L. (2006). *Menopause: A Biocultural Perspective*. New Brunswick, New Jersey: Rutgers University Press.

Leon, D. A., Chenet, L., Shkolnikov, V. M. *et al.* (1997). Huge variation in Russian mortality rates 1984–94: artefact, alcohol, or what? *The Lancet*, **350**, 383–8.

Leonard, W. R. (2000). Human nutritional evolution. In *Human Biology: An Evolutionary and Biocultural Perspective*, ed. S. Stinson, B. Bogin, R. Huss-Ashmore and D. O'Rourke. New York: Wiley-Liss, pp. 295–343.

Lett, H. S., Blumenthal, J. A., Babyak, M. A. *et al.* (2004). Depression as a risk factor for coronary artery disease: evidence, mechanisms and treatment. *Psychosomatic Medicine*, **66**, 305–15.

Lewontin, R. C. (1982). *Human Diversity*. New York: Scientific American Library.

Li, C., Ford, E. S., Mokdad, A. H. and Cook, S. (2006). Recent trends in waist circumference and waist–height ratio among US children and adolescents. *Pediatrics*, **118**, e1390–8.

Lieberman, L. S. (2003). Dietary, evolutionary, and modernizing influences on the prevalence of type 2 diabetes. *Annual Review of Nutrition*, **23**, 345–77.

Linden, W., Stossel, C. and Maurice, J. (1996). Psychosocial interventions for patients with coronary artery disease: a meta-analysis. *Archives of Internal Medicine*, **156**, 745–52.

Lipson, S. F. (2001). Metabolism, maturation, and ovarian function. In *Reproductive Ecology and Human Evolution*, ed. P. T. Ellison. New York: Aldine de Gruyter, pp. 235–48.

Lipson, S. F. and Ellison, P. T. (1996). Comparison of salivary steroid profiles in naturally occurring conception and non-conception cycles. *Human Reproduction*, **11**, 2090–6.

Liu, S., Willett, W. C., Manson, J. E., Hu, F. B., Rosner, B. and Colditz, G. (2003). Relation between changes in intakes of dietary fiber and grain products and changes in weight and development of obesity among middle-aged women. *American Journal of Clinical Nutrition*, **78**, 920–7.

Livingston, C. and Collison, M. (2002). Sex steroids and insulin resistance. *Clinical Science*, **102**, 151–66.

Lock, M. and Kaufert, P. (2001). Menopause, local biologies, and cultures of aging. *American Journal of Human Biology*, **13**, 494–504.

Ludwig, D. S. (2002). The glycemic index: physiological mechanisms relating to obesity, diabetes, and cardiovascular disease. *Journal of the American Medical Association*, **287**, 2414–23.

Ludwig, D. S., Peterson, K. E. and Gortmaker, S. L. (2001). Relation between consumption of sugar-sweetened drinks and childhood obesity: a prospective, observational analysis. *The Lancet*, **357**, 505–8.

Macleod, J. and Davey Smith, G. (2003). Psychosocial factors and public health: a suitable case for treatment. *Journal of Epidemiology and Community Health*, **57**, 565–70.

Maizels, R. M. (2005). Infections and allergy – helminths, hygiene and host immune regulation. *Current Opinion in Immunology*, **17**, 656–61.

Mann, J. (2004). Free sugars and human health: sufficient evidence for action? *The Lancet*, **363**, 1068–70.

Mann, J. I. (2002). Diet and risk of coronary heart disease and type 2 diabetes. *The Lancet*, **360**, 783–9.

Marmot, M., Shipley, M. and Rose, G. (1984). Inequalities in death – specific explanations of a general pattern? *The Lancet*, **i**, 1003–6.

Marshall, C., Hitman, G. A., Partridge, C. J. *et al.* (2005). Evidence that an isoform of Calpain-10 is a regulator of exocytosis in pancreatic beta-cells. *Molecular Endocrinology*, **19**, 213–24.

Martin, R. M., Gunnell, D. and Davey Smith, G. (2005). Breastfeeding in infancy and blood pressure in later life: systematic review and meta-analysis. *American Journal of Epidemiology*, **161**, 15–26.

Martinez, F. D. (1994). Role of viral infections in the inception of asthma: could they be protective? *Thorax*, **49**, 1189–91.

Martorell, R., Khan, L. K., Hughes, M. L. and Grummer-Strawn, L. M. (1998). Obesity in Latin American women and children. *Journal of Nutrition*, **128**, 1464–73.

Mascie-Taylor, C. G. N. (1993). The biological anthropology of disease. In *The Anthropology of Disease*, ed. C. G. N. Mascie-Taylor. Oxford: Oxford University Press, pp. 1–72.

Maskarinec, G., Franke, A., Williams, A. *et al.* (2004). Effects of a 2-year randomized soy intervention on sex hormone levels in premenopausal women. *Cancer Epidemiology, Biomarkers and Prevention*, **13**, 1736–44.

Masoli, M., Fabian, D., Holt, S. and Beasley, R. for the Global Initiative for Asthma (GINA) Program (2004a). The global burden of asthma: executive summary of the GINA Dissemination Committee Report. *Allergy*, **59**, 469–78.

Masoli, M., Fabian, D., Holt, S. H. A. and Beasley, R. (2004b). *Global Burden of Asthma*: Global Initiative for Asthma.

Mathers, C. D. and Loncar, D. (2006). Projections of global mortality and burden of disease from 2002 to 2030. *PLoS Medicine*, **3**, 2011–30.

Matricardi, P. M., Bjorksten, B., Bonini, S. *et al.* for the EAACI Task Force 7 (2003). Microbial products in allergy prevention and therapy. *Allergy*, **58**, 461–71.

Matsuda, K., Nishi, Y., Okamatsu, Y., Kojima, M. and Matsuishi, T. (2006). Ghrelin and leptin: a link between obesity and allergy? *Journal of Allergy and Clinical Immunology*, **117**, 705–6.

Matthews, K. A. (2005). Psychological perspectives on the development of coronary heart disease. *American Psychologist*, **60**, 783–96.

Matthews, K. A., Raikkonen, K., Everson, S. A. *et al.* (2000). Do the daily experiences of healthy men and women vary according to occupational prestige and work strain? *Psychosomatic Medicine*, **62**, 346–53.

Mayor, S. (2005). 23% of babies in England are delivered by caesarean section. *British Medical Journal*, **330**, 806.

McDaniel, C. N. and Gowdy, J. M. (2000). *Paradise for Sale: A Parable of Nature*. Berkeley, California: University of California Press.

McElroy, A. and Townsend, P. K. (2004). *Medical Anthropology in Ecological Perspective*. Boulder, Colorado: Westview Press.

McEwen, B. S. and Stellar, E. (1993). Stress and the individual – mechanisms leading to disease. *Archives of Internal Medicine*, **153**, 2093–110.

McEwen, B. S. and Wingfield, J. C. (2003). The concept of allostatis in biology and biomedicine. *Hormones and Behavior*, **43**, 2–15.

McGarvey, S. T., Bindon, J. R., Crews, D. E. and Schendel, D. E. (1989). Modernization and adiposity: causes and consequences. In *Human Population Biology*, ed. M. A. Little and J. D. Haas. Oxford: Oxford University Press, pp. 263–80.

McGee, R., Williams, S. and Elwood, M. (1994). Depression and the development of cancer – a meta-analysis. *Social Science and Medicine*, **1994**, 187–92.

McKeigue, P. M. (1996). Metabolic consequences of obesity and body fat pattern: lessons from migrant studies. In *The Origins and Consequences of Obesity. CIBA Foundation Symposium 201*, ed. D. J. Chadwick and G. Cardew. Chichester: John Wiley, pp. 54–63.

McKeigue, P. M., Shah, B. and Marmot, M. G. (1991). Relation of central obesity and insulin resistance with high diabetes prevalence and cardiovascular risk in South Asians. *The Lancet*, **337**, 382–6.

McKenna, J. J., Mosko, S. and Richard, C. (1999). Breast-feeding and mother–infant cosleeping in relation to SIDS prevention. In *Evolutionary Medicine*, ed. W. R. Trevathan, E. O. Smith and J. J. McKenna. New York: Oxford University Press, pp. 53–74.

McMichael, A. J. (2001). *Human Frontiers, Environments and Disease*. Cambridge: Cambridge University Press.

McPherson, K., Steel, C. M. and Dixon, J. M. (2000). Breast cancer – epidemiology, risk factors, and genetics. *British Medical Journal*, **321**, 624–8.

Meek, J. Y. (2001). Breastfeeding in the workplace. *Pediatric Clinics of North America*, **48**, 461–74.

Menacker, F. (2005). *Trends in Cesarean Rates for First Births and Repeat Cesarean Rates for Low Risk Women: United States, 1990–2000. National Vital Statistics Report 54*: 1–12. Hyattsville, Maryland: National Center for Health Statistics.

Mendelsohn, M. E. and Karas, R. H. (1999). The protective effects of estrogen on the cardiovascular system. *New England Journal of Medicine*, **340**, 1801–11.

Meyer, V. F. (2001). The medicalization of the menopause: Critique and consequences. *International Journal of Health Services*, **31**, 769–92.

Midthjell, K., Kruger, O., Holmen, J. *et al.* (1999). Rapid changes in the prevalence of obesity and known diabetes in an adult Norwegian population – the Nord-Trondelag Health Surveys: 1984–1986 and 1995–1997. *Diabetes Care*, **22**, 1813–20.

Miller, W. C., Koceja, D. M. and Hamilton, E. J. (1997). A meta-analysis of the past 25 years of weight loss research using diet, exercise or diet plus exercise intervention. *International Journal of Obesity and Related Metabolic Disorders*, **21**, 941–47.

Misra, A. and Vikram, N. K. (2004). Insulin resistance syndrome (metabolic syndrome) and obesity in Asian Indians: evidence and implications. *Nutrition*, **20**, 482–91.

Mokdad, A. H., Serdula, M. K., Dietz, W. H., Bowman, B. A., Marks, J. S. and Koplan, J. P. (1999). The spread of the obesity epidemic in the United States, 1991–1998. *Journal of the American Medical Association*, **282**, 1519–22.

Moore, S. E., Halsall, I., Howarth, D., Poskitt, E. M. E. and Prentice, A. M. (2001). Glucose, insulin and lipid metabolism in rural Gambians exposed to early malnutrition. *Diabetic Medicine*, **18**, 645–53.

Muehlenbein, M. P. and Bribiescas, R. G. (2005). Testosterone-mediated immune functions and male life histories. *American Journal of Human Biology*, **17**, 527–58.

Mufunda, J., Chatora, R., Ndambukuwa, Y. *et al.* (2006). Emerging non-communicable disease epidemic in Africa: preventive measures from the WHO Regional Office for Africa. *Ethnicity and Disease*, **16**, 521–6.

Murch, S. H. (2001). Toll of allergy reduced by probiotics. *The Lancet*, **357**, 1057–9.

Murray, C. J. L. and Lopez, A. D. (1997). Alternative projections of mortality and disability by cause 1990–2020: global burden of disease study. *The Lancet*, **349**, 1498–504.

Murray, C. J. L., Lauer, J. A., Hutubessy, R. C. W. *et al.* (2003). Effectiveness and costs of interventions to lower systolic blood pressure and cholesterol: a global and regional analysis on reduction of cardiovascular-disease risk. *The Lancet*, **361**, 717–25.

Muscat, J. E., Harris, R. E., Haley, N. J. and Wynder, E. L. (1991). Cigarette smoking and plasma cholesterol. *American Heart Journal*, **121**, 141–7.

Mussa, F. F., Chai, H., Wang, X., Yao, Q., Lumsden, A. B. and Chen, C. (2006). *Chlamydia pneumoniae* and vascular disease: an update. *Journal of Vascular Surgery*, **43**, 1301–7.

Must, A. and Colclough-Douglas, S. (2001). Adult health sequelae of pediatric obesity. In *Obesity, Growth and Development*, ed. F. E. Johnston and G. D. Foster. London: Smith-Gordon, pp. 185–98.

Must, A., Spadano, J., Coakley, E. H., Field, A. E., Colditz, G. and Dietz, W. H. (1999). The disease burden associated with overweight and obesity. *Journal of the American Medical Association*, **282**, 1523–9.

Muti, P. (2004). The role of endogenous hormones in the etiology and prevention of breast cancer: the epidemiological evidence. *Annals of the New York Academy of Sciences*, **1028**, 273–182.

National Sleep Foundation (2002). *"Sleep in America" Poll*. Washington, DC: National Sleep Foundation.

Nazroo, J. (2003). The structuring of ethnic inequalities in health: economic position, racial discrimination, and racism. *American Journal of Public Health*, **93**, 277–84.

Neel, J. V. (1962). Diabetes mellitus: a "thrifty" genotype rendered detrimental by "progress"? *American Journal of Human Genetics*, **14**, 353–62.

Neel, J. V. (1982). The thrifty genotype revisited. In *The Genetics of Diabetes Mellitus*, ed. J. Köbberling and R. Tattersall. Amsterdam: Academic Press, pp. 137–47.

Neel, J. V., Weder, A. B. and Julius, S. (1998). Type II diabetes, essential hypertension, and obesity as 'syndromes of impaired genetic homeostasis': the 'thrifty genotype' hypothesis enters the 21st century. *Perspectives in Biology and Medicine*, **42**, 44–74.

Ness, R. B., Haggerty, C. L., Harger, G. and Ferrell, R. E. (2004). Differential distribution of allelic variants in cytokine genes among African Americans and white Americans. *American Journal of Epidemiology*, **160**, 1033–8.

Ness, R. B. (2000). Is depression an adaptation? *Archives of General Psychiatry*, **57**, 14–20.

Nesse, R. M. (2004). Natural selection and the elusiveness of happiness. *Philosophical Transactions of the Royal Society of London B*, **359**, 1333–47.

Nesse, R. M. and Williams, G. C. (1994). *Evolution and Healing: The New Science of Darwinian Medicine*. London: Phoenix.

Nesse, R. M., Stearns, S. C. and Omenn, G. S. (2006). Medicine needs evolution. *Science*, **311**, 1071.

Nettle, D. (2004). Evolutionary origins of depression: a review and reformulation. *Journal of Affective Disorders*, **81**, 91–102.

Netuveli, G., Hurwitz, B., Levy, M. *et al.* (2005). Ethnic variations in UK asthma frequency, morbidity, and health-service use: a systematic review and meta-analysis. *The Lancet*, **365**, 312–17.

Neuhausen, S. L. (1999). Ethnic differences in cancer risk resulting from genetic variation. *Cancer*, **86**, 2575–82.

Newsome, C. A., Shiell, A. W., Fall, C. H. D., Phillips, D. I. W., Shier, R. and Law, C. M. (2003). Is birth weight related to later glucose and insulin metabolism? A systematic review. *Diabetic Medicine*, **20**, 339–48.

Nielsen, S. J. and Popkin, B. M. (2003). Patterns and trends in food portion sizes, 1977–1998. *Journal of the American Medical Association*, **289**, 450–3.

NIH State-of-the–Science Panel (2005). National Institutes of Health State-of-the–Science conference statement: management of menopause-related symptoms. *Annals of Internal Medicine*, **142**, 1003–13.

Núñez-de la Mora, A., Chatterton, R. T., Choudhury, O. A., Napolitano, D. A. and Bentley, G. R. (2007). Childhood conditions influence adult progesterone levels. *PLoS Medicine*, doi: 10.1371/journal. pmed. 0040167.

O'Dea, K. (1984). Marked improvement in carbohydrate and lipid metabolism in diabetic Australian Aborigines after temporary reversion to traditional lifestyle. *Diabetes*, **33**, 596–603.

O'Dea, K. (1991). Traditional diet and food preferences of Australian Aboriginal hunter–gatherers. *Philosophical Transactions of the Royal Society of London B*, **334**, 233–41.

O'Dea, K. and Piers, L. (2002). Diabetes. In *The Nutrition Transition: Diet and Disease in the Developing World*, ed. B. Caballero and B. M. Popkin. Amsterdam: Elsevier, pp. 165–90.

O'Dea, K., Hopper, J., Patel, M., Traianedes, K. and Kubisch, D. (1993). Obesity, diabetes, and hyperlipidemia in a Central Australian Aboriginal community with a long history of acculturation. *Diabetes Care*, **16**, 1004–10.

O'Sullivan, A. J., Martin, A. and Brown, M. A. (2001). Efficient fat storage in premenopausal women and in early pregnancy: a role for estrogen. *Journal of Clinical Endocrinology and Metabolism*, **86**, 4951–6.

Oken, E. and Gillman, M. W. (2003). Fetal origins of obesity. *Obesity Research*, **11**, 496–506.

Omran, A. R. (1971). The epidemiologic transition: a theory of the epidemiology of population change. *Milbank Memorial Fund Quarterly*, **49**, 509–38.

Ong, K. K. and Dunger, D. B. (2004). Birth weight, infant growth and insulin resistance. *European Journal of Endocrinology*, **151**, U131–U139.

Ong, K. L. and Dunger, D. B. (2000). Thrifty genotypes and phenotypes in the pathogenesis of type 2 diabetes mellitus. *Journal of Pediatric Endocrinology and Metabolism*, **13**, 1419–24.

Ongphiphadhanakul, B., Rajatanavin, R., Chailurkit, L. *et al.* (1995). Serum testosterone and its relation to bone mineral density and body composition in normal males. *Clinical Endocrinology*, **43**, 727–33.

Orth-Gomer, K., Eriksson, I., Moser, V., Theorell, T. and Fredlund, P. (1994). Lipid lowering through work stress reduction. *International Journal of Behavioral Medicine*, **1**, 204–14.

Osei, K., Schuster, D. P., Owusu, S. K. and Amoah, A. G. B. (1997). Race and ethnicity determine serum insulin and C-peptide concentrations and hepatic insulin extraction and insulin clearance: comparative studies of three populations of west African ancestry and white Americans. *Metabolism*, **46**, 53–8.

Ostler, K., Thompson, C., Kinmonth, A.-L. K., Peveler, R. C. and Stevens, L. (2001). Influence of socio-economic deprivation on the prevalence and outcome of depression in primary care. *British Journal of Psychiatry*, **178**, 12–17.

Owen, C. G., Whincup, P. H., Odoki, K., Gilg, J. A. and Cook, D. G. (2002). Infant feeding and blood cholesterol: a study in adolescents and a systematic review. *Pediatrics*, **110**, 597–608.

Owens, J. F., Stoney, C. M. and Matthews, K. A. (1993). Menopausal status influences ambulatory blood pressure levels and blood pressure changes during mental stress. *Circulation*, **88**, 2794–802.

Palacios, S. (1999). Current perspectives on the benefits of HRT in menopausal women. *Maturitas*, **33**, S1–S13.

Pan, X.-R., Li, G.-W., Hu, Y. H. *et al.* (1997). Effects of diet and exercise in preventing NIDDM in people with impaired glucose tolerance. The Da Qing IGT and diabetes study. *Diabetes Care*, **20**, 537–44.

Paradies, Y. C., Montoya, M. J. and Fullerton, S. M. (2007). Racialised genetics and the study of complex diseases: the thrifty genotype revisited. *Perspectives in Biology and Medicine*, **50**, 203–27.

Parkin, D. M. and Fernández, L. M. G. (2006). Use of statistics to assess the global burden of breast cancer. *Breast Journal*, **12** (Suppl. 1), S70–80.

Parkin, D. M., Bray, F., Ferlay, J. and Pisani, P. (2005). Global cancer statistics, 2002. *CA: A Cancer Journal for Clinicians*, **55**, 74–108.

Parnia, S., Borwn, J. L. and Frew, A. J. (2002). The role of pollutants in allergic sensitization and the development of asthma. *Allergy*, **57**, 1111–17.

Pasquali, R. (2006). Obesity, fat distribution and infertility. *Maturitas*, **54**, 363–71.

Pasquali, R., Pelusi, C., Genghini, S., Cacciari, M. and Gambineri, A. (2003). Obesity and reproductive disorders in women. *Human Reproduction Update*, **9**, 359–72.

Patel, S. M. and Nestler, J. E. (2006). Fertility in polycystic ovary syndrome. *Endocrinology and Metabolism Clinics of North America*, **35**, 137–55.

Patz, J. A., Epstein, P. R., Burke, T. A. and Balbus, J. M. (1996). Global climate change and emerging infectious diseases. *Journal of the American Medical Association*, **275**, 217–23.

Pearce-Duvet, J. M. C. (2006). The origin of human pathogens: evaluating the role of agriculture and domestic animals in the evolution of human disease. *Biological Reviews*, **81**, 369–82.

Penders, J., Thijs, C., Vink, C. *et al.* (2006). Factors influencing the composition of the intestinal microbiota in early infancy. *Pediatrics*, **118**, 511–21.

Pérez-Perdomo, R., Pérez-Cardona, C., Disdier-Flores, O. and Cintrón, Y. (2003). Prevalence and correlates of asthma in the Puerto Rican population: Behavioral Risk Factor Surveillance System, 2000. *Journal of Asthma*, **40**, 465–74.

Pettitt, D. J., Aleck, K., Baird, H., Carraher, M., Bennett, P. and Knowler, W. C. (1988). Congenital susceptibility to NIDDM: role of intrauterine environment. *Diabetes*, **37**, 622–8.

Pettitt, D. J., Forman, M. R., Hanson, R. L., Knowler, W. C. and Bennett, P. H. (1997). Breastfeeding and incidence of non-insulin-dependent diabetes mellitus in Pima Indians. *The Lancet*, **350**, 166–8.

Pi-Sunyer, F. X. (2002). Glycemic index and disease. *American Journal of Clinical Nutrition*, **76** (Suppl.), 290S–8S.

Pirart, J. (1978). Diabetes mellitus and its degenerative complications: a prospective study of 4,400 patients observed between 1947 and 1973 (Part 1). *Diabetes Care*, **1**, 168–88.

Platts-Mills, T. A. E., Vervloet, D., Thomas, W. R., Aalberse, R. C. and Chapman, M. D. (1997). Indoor allergens and asthma: report of the Third International Workshop. *Journal of Allergy and Clinical Immunology*, **100**, S2–24.

Platz, E. and Giovannucci, E. (2004). The epidemiology of sex steroid hormones and their signaling and metabolic pathways in the etiology of prostate cancer. *Journal of Steroid Biochemistry and Molecular Biology*, **92**, 237–53.

Pollard, I. (1994). *A Guide to Reproduction: Social Issues and Human Concerns*. Cambridge: Cambridge University Press.

Pollard, T. M. (1997). Environmental change and cardiovascular disease: a new complexity. *Yearbook of Physical Anthropology*, **40**, 1–24.

Pollard, T. M. (2000). Adrenaline. In *Encyclopedia of Stress*, ed. G. Fink. San Diego: Academic Press, pp. 52–8.

Pollard, T. M. and Ice, G. H. (2006). Measuring hormonal variation in the hypothalamic pituitary adrenal axis: cortisol. In *Measuring Stress in Humans: A Practical Guide for the Field*, ed. G. H. Ice and G. D. James. Cambridge: Cambridge University Press, pp. 122–57.

Pollard, T. M. and Unwin, N. C. (2007). Impaired reproductive function in western and 'westernising' populations: an evolutionary approach. In *New Perspectives in Evolutionary Medicine*, ed. W. R. Trevathan, E. O. Smith and J. J. McKenna. Oxford: Oxford University Press.

Pollard, T. M., Brush, G. and Harrison, G. A. (1991). Geographic distribution of within-population variability in blood pressure. *Human Biology*, **63**, 643–61.

Pollard, T. M., Núñez-de la Mora, A. and Unwin, N. C. (in press). Evolutionary perspectives on type 2 diabetes in Asia. In *Medicine and Evolution: Current Applications, Future Prospects*, ed. S. Elton and P. O'Higgins. Boca Raton: Taylor and Francis.

Pollard, T. M., Ungpakorn, G., Harrison, G. A. and Parkes, K. R. (1996). Epinephrine and cortisol responses to work: a test of the models of Frankenhaeuser and Karasek. *Annals of Behavioral Medicine*, **18**, 229–37.

Pollock, K. (1988). On the nature of social stress: production of a modern mythology. *Social Science and Medicine*, **26**, 381–92.

Pond, C. M. (1998). *The Fats of Life*. Cambridge: Cambridge University Press.

Popham, F. and Mitchell, R. (2006). Leisure time exercise and personal circumstances in the working age population: longitudinal analysis of the British household panel survey. *Journal of Epidemiology and Community Health*, **60**, 270–4.

Popkin, B. M. (1999). Urbanization, lifestyle changes and the nutrition transition. *World Development*, **27**, 1905–16.

Popkin, B. M. (2001). The nutrition transition and obesity in the developing world. *Journal of Nutrition*, **131**, 871S–3S.

Popkin, B. M. and Gordon-Larsen, P. (2004). The nutrition transition: worldwide obesity dynamics and their determinants. *International Journal of Obesity*, **28**, S2–9.

Poretsky, L., Cataldo, N. A., Rosenwaks, Z. and Giudice, L. C. (1999). The insulin-related ovarian regulatory system in health and disease. *Endocrine Reviews*, **20**, 535–82.

Portengen, L., Preller, L., Tielen, M., Doekes, G. and Heederik, D. (2005). Endotoxin exposure and atopic sensitization in adult pig farmers. *Journal of Allergy and Clinical Immunology*, **115**, 797–802.

Potter, L. B., Rogler, L. H. and Moscicki, E. K. (1995). Depression among Puerto Ricans in New York City – the Hispanic Health and Nutrition Examination Survey. *Social Psychiatry and Psychiatric Epidemiology*, **30**, 185–93.

Poulain, M., Doucet, M., Major, G. C. *et al.* (2006). The effect of obesity on chronic respiratory diseases: pathophysiology and therapeutic strategies. *Canadian Medical Association Journal*, **174**, 1293–9.

Prentice, A. M. (2001). Fires of life: the struggles of an ancient metabolism in a modern world. *BNF Nutrition Bulletin*, **26**, 13–27.

Prentice, A. M. and Jebb, S. A. (1995). Obesity in Britain: gluttony or sloth. *British Medical Journal*, **311**, 437–9.

Prentice, A. M., Rayco-Solon, P. and Moore, S. E. (2005). Insights from the developing world: thrifty genotypes and thrifty phenotypes. *Proceedings of the Nutrition Society*, **64**, 153–61.

Preston, S. H. (1976). *Mortality Patterns in National Populations: With Special Reference to Recorded Causes of Death*. New York: Academic Press.

Price, J. F. and Fowkes, G. R. (1997). Risk factors and the sex differential in coronary artery disease. *Epidemiology*, **8**, 584–91.

Prior, I. A. M. (1971). The price of civilization. *Nutrition Today*, **6**, 2–11.

Raben, N., Barbetti, F., Cama, A. *et al.* (1991). Normal coding sequence of insulin gene in Pima Indians and Nauruans, two groups with highest prevalence of type II diabetes. *Diabetes*, **40**, 118–22.

Rajkumar, L., Guzman, R. C., Yang, J., Thordarson, G., Talamantes, F. and Nandi, S. (2004). Prevention of mammary carcinogenesis by short-term estrogen and progestin treatments. *Breast Cancer Research*, **6**, R31–7.

Randolph, J. F., Sowers, M., Gold, E. B. *et al.* (2003). Reproductive hormones in the early menopausal transition: relationship to ethnicity, body size, and menopausal status. *Journal of Clinical Endocrinology and Metabolism*, **88**, 1516–22.

Ravelli, A. C. J., van der Meulen, J. H. P., Osmond, C., Barker, D. J. P. and Bleker, O. P. (2000). Infant feeding and adult glucose tolerance, lipid profile, blood pressure, and obesity. *Archives of Disease in Childhood*, **82**, 248–52.

Ray, C. and Stevens, J. R. (1995). *Sacred Legends*. Manotick, ON: Penumbra Press.

Reaven, G. M. (1988). Role of insulin resistance in human disease. *Diabetes*, **37**, 1595–607.

Rebuffé-Scrive, M., Enk, L., Crona, N. *et al.* (1985). Fat cell metabolism in different regions in women: effect of menstrual cycle, pregnancy, and lactation. *Journal of Clinical Investigation*, **75**, 1973–6.

Reilly, J. J., Jackson, D. M., Montgomery, C. *et al.* (2004). Total energy expenditure and physical activity in young Scottish children: mixed longitudinal study. *The Lancet*, **363**, 211–12.

Reilly, J. J., Armstrong, J., Dorosty, A. R. *et al.* for the Avon Longitudinal Study of Parents and Children Study Team (2005). Early life risk factors for obesity in childhood: cohort study. *British Medical Journal*, **330**, 1357–9.

Reinhard, K. J. (1988). Cultural ecology of prehistoric parasitism on the Colorado Plateau as evidenced by coprology. *American Journal of Physical Anthropology*, **77**, 355–66.

Relethford, J. H. (1994). *Fundamental of Biological Anthropology*. Mountain View, California: Mayfield Publishing Company.

Renehan, A. G., Zwahlen, M., Minder, C., O'Dwyer, S. T., Shalet, S. M. and Egger, M. (2004). Insulin-like growth factor (IGF)-1, IGF binding protein-3, and cancer risk: systematic review and meta-regression analysis. *The Lancet*, **363**, 1346–53.

Richards, M. P. (2002). A brief review of the archaeological evidence for Palaeolithic and Neolithic subsistence. *European Journal of Clinical Nutrition*, **56**, 1–9.

Ridker, P. M. (2002). On evolutionary biology, inflammation, infection and the causes of atherosclerosis. *Circulation*, **105**, 2–4.

Riedler, J., Braun-Fahrlander, C., Eder, W. *et al.* and the ALEX study team (2001). Exposure to farming in early life and development of asthma and allergy: a cross-sectional survey. *The Lancet*, **358**, 1129–33.

Rissanen, A. M., Heliovaara, M., Knekt, P., Reunanen, A. and Aromaa, A. (1991). Determinants of weight gain and overweight in adult Finns. *European Journal of Clinical Nutrition*, **45**, 419–30.

Ritenbaugh, C. and Goodby, C. S. (1989). Beyond the thrifty gene: metabolic implications of prehistoric migration into the new world. *Medical Anthropology*, **11**, 227–36.

Roberts, C. A. and Cox, M. (2003). *Health and Disease in Britain*. Stroud: Sutton Publishing.

Roberts, E. M. (2002). Racial and ethnic disparities in childhood asthma diagnosis: the role of clinical findings. *Journal of the National Medical Association*, **94**, 215–23.

Robinson, T. N. (2001). Population-based obesity prevention for children and adolescents. In *Obesity, Growth and Development*, ed. F. E. Johnston and G. D. Foster. London: Smith–Gordon, pp. 129–41.

Rode, A. and Shephard, R. J. (1971). Cardiorespiratory fitness of an Arctic community. *Journal of Applied Physiology*, **31**, 519–26.

Rodriguez, M. A., Winkleby, M. A., Ahn, D., Sundquist, J. and Kraemer, H. C. (2002). Identification of population subgroups of children and adolescents with high asthma prevalence. *Archives of Pediatric and Adolescent Medicine*, **156**, 269–75.

Rook, G. A. W., Adams, V., Hunt, J., Palmer, R., Martinelli, R. and Rosa Brunet, L. (2004). Mycobacteria and other environmental organisms as immunomodulators for immunoregulatory disorders. *Springer Seminars in Immunopathology*, **25**, 237–55.

Roper, N. A., Bilous, R. W., Kelly, W. F., Unwin, N. C. and Connolly, V. M. (2001). Excess mortality in a population with diabetes and the impact of material deprivation: longitudinal, population based study. *British Medical Journal*, **322**, 1389–93.

Rose, D., Mannino, D. M. and Leaderer, B. P. (2006). Asthma prevalence among US adults, 1998–2000: role of Puerto Rican ethnicity and behavioral and geographic factors. *American Journal of Public Health*, **96**, 880–8.

Rosenblatt, K. A., Thomas, D. B. and The WHO Collaborative Study of Neoplasia and Steroid Contraceptives (1995). Prolonged lactation and endometrial cancer. *International Journal of Epidemiology*, **24**, 499–503.

Ross, R. K., Coetzee, G. A., Reichardt, J., Skinner, E. and Henderson, B. E. (1995). Does the racial-ethnic variation in prostate cancer risk have a hormonal basis? *Cancer*, **75**, 1778–82.

Roumain, J., Charles, M. A., de Courten, M. P. *et al.* (1998). The relationship of menstrual irregularity to Type 2 diabetes in Pima Indian women. *Diabetes Care*, **21**, 346–9.

Royal, C. D. M. and Dunston, G. M. (2004). Changing the paradigm from 'race' to human genome variation. *Nature Genetics*, **36**, S5–7.

Runciman, W. G. (2005). Stone age sociology. *Journal of the Royal Anthropological Institute*, **11**, 129–42.

Ryan, A. S., Wenjun, Z. and Acosta, A. (2002). Breastfeeding continues to increase into the new millenium. *Pediatrics*, **110**, 1103–9.

Sackett, R. D. (1996). *Time, Energy, and the Indolent Savage*. Unpublished PhD thesis. University of California, Los Angeles.

Saelens, B. E., Sallis, J. F., Black, J. B. and Chen, D. (2003). Neighbourhood-based differences in physical activity: an environment scale evaluation. *American Journal of Public Health*, **93**, 1552–8.

Sahlins, M. (1972). *Stone Age Economics*. New York: Aldine de Gruyter.

Sánchez-Castillo, C. P., Velásquez-Monroy, O., Lara-Esqueda, A. *et al.* (2005). Diabetes and hypertension increases in a society with abdominal obesity: results of the Mexican National Health Survey 2000. *Public Health Nutrition*, **8**, 53–60.

Sapolsky, R. M., Krey, L. C. and McEwen, B. S. (1986). The neuroendocrinology of stress and aging: the glucocorticoid cascade hypothesis. *Endocrine Reviews*, **7**, 284–301.

Sapolsky, R. M., Romero, L. M. and Munck, A. U. (2000). How do glucocorticoids influence stress responses? Integrating permissive, suppressive, stimulatory, and preparative actions. *Endocrine Reviews*, **21**, 55–89.

Saxton, J. M. (2006). Diet, physical activity and energy balance and their impact on breast and prostate cancers. *Nutrition Research Reviews*, **19**, 197–215.

Schaub, B., Lauener, R. and von Mutius, E. (2006). The many faces of the hygiene hypothesis. *Journal of Allergy and Clinical Immunology*, **117**, 969–77.

Schieffelin, E. L. (1985). The cultural analysis of depressive affect: an example from New Guinea. In *Culture and Depression: Studies in the Anthropology and Cross-Cultural Psychiatry of Affect and Disorder*, ed. A. Kleinman and B. Good. Berkeley: University of California Press, pp. 101–33.

Schofield, R. and Reher, D. (1991). The decline of mortality in Europe. In *The Decline of Mortality in Europe*, ed. R. Schofield, D. Reher and D. Bideau. Oxford: Clarendon, pp. 1–17.

Schröder, F. H. (1996). Impact of ethnic, nutritional and environmental factors on prostate cancer. In *Pharmacology, Biology, and Clinical Applications of Androgens*, ed. S. Bhasin, H. L. Gabelnick, J. M. Spieleret al. New York: Wiley-Liss, pp. 121–36.

Schulz, L. O., Bennett, P. H., Ravussin, E. *et al.* (2006). Effects of traditional and western environments on prevalence of type 2 diabetes in Pima Indians in Mexico and the US. *Diabetes Care*, **29**, 1866–71.

Schulze, M. B., Manson, J. E., Ludwig, D. S. *et al.* (2004). Sugar-sweetened beverages, weight gain, and incidence of type 2 diabetes in young and middle-aged women. *Journal of the American Medical Association*, **292**, 927–34.

Schutz, Y., Weinsier, R. L. and Hunter, G. R. (2001). Assessment of free-living physical activity in humans: an overview of currently available and proposed new measures. *Obesity Research*, **9**, 368–79.

Scott, S. and Duncan, C. J. (2001). *Biology of Plagues: Evidence from Historical Populations*. Cambridge: Cambridge University Press.

Scragg, R. K. R., Fraser, A. and Metcalf, P. A. (1996). *Helicobacter pylori* seropositivity and cardiovascular risk factors in a multicultural workforce. *Journal of Epidemiology and Community Health*, **50**, 578–9.

Scrivener, S., Yemaneberhan, H., Zebenigus, M. *et al.* (2001). Independent effects of intestinal parasite infection and domestic allergen exposure on risk of wheeze in Ethiopia: a nested case-control study. *The Lancet*, **358**, 1493–9.

Seale, C. (2000). Changing patterns of death and dying. *Social Science and Medicine*, **51**, 917–30.

Seidell, J. C. (1995). Obesity in Europe: scaling an epidemic. *International Journal of Obesity*, **19** (Suppl. 3), S1–4.

Seidell, J. C. (2000). Obesity, insulin resistance and diabetes – a worldwide epidemic. *British Journal of Nutrition*, **83** (Suppl. 1), S5–8.

Sellen, D. W. and Smay, D. B. (2001). Relationship between subsistence and age at weaning in "preindustrial" societies. *Human Nature*, **12**, 47–87.

Sellers, E. A. C., Triggs-Raine, B., Rockman-Greenberg, C. and Dean, H. J. (2002). The prevalence of the HNF-1alpha G319S mutation in Canadian Aboriginal youth with type 2 diabetes. *Diabetes Care*, **25**, 2202–6.

Sephton, S. E., Sapolsky, R. M., Kraemer, H. C. and Spiegel, D. (2000). Diurnal cortisol rhythm as a predictor of breast cancer survival. *Journal of the National Cancer Institute*, **92**, 994–1000.

Shaneyfelt, T., Husein, R., Bubley, G. and Mantzoros, C. S. (2000). Hormonal predictors of prostate cancer: a meta-analysis. *Journal of Clinical Oncology*, **18**, 847–53.

Shanley, D. P. and Kirkwood, T. B. L. (2001). Evolution of the human menopause. *Bioessays*, **23**, 282–7.

Sharpe, R. M. and Franks, S. (2002). Environment, lifestyle and infertility – an inter-generational issue. *Nature Medicine*, **8** (Suppl. 11), 33–40.

Sharpe, R. M. and Skakkebaek, N. E. (1993). Are oestrogens involved in falling sperm counts and disorders of the male reproductive tract? *The Lancet*, **341**, 1392–5.

Shea, J. L. (2006). Parsing the ageing Asian woman: symptom results from the China Study of Midlife Women. *Maturitas*, **55**, 36–50.

Sherry, D. S. and Marlowe, F. W. (2007). Anthropometric data indicate nutritional homogeneity in Hadza Foragers of Tanzania. *American Journal of Human Biology*, **19**, 107–18.

Shetty, P. S., Henry, C. J. K., Black, A. E. and Prentice, A. M. (1996). Energy requirements of adults: an update on basal metabolic rates (BMRs) and physical activity levels (PALs). *European Journal of Clinical Nutrition*, **50** (Suppl. 1), S11–23.

Shimuzu, H., Ross, R. K., Bernstein, L., Pike, M. C. and Henderson, B. E. (1990). Serum oestrogen levels in postmenopausal women: comparison of American whites and Japanese in Japan. *British Journal of Cancer*, **62**, 451–3.

Shore, S. A. (2006). Obesity and asthma: cause for concern. *Current Opinion in Pharmacology*, **6**, 230–6.

Shusterman, D. J., Murphy, M. A. and Balmes, J. R. (1998). Subjects with seasonal allergic rhinitis and nonrhinitic subjects react differently to nasal provocation with chlorine gas. *Journal of Allergy and Clinical Immunology*, **101**, 732–40.

Sicherer, S. H., Muñoz-Furlong, A. and Sampson, H. A. (2003). Prevalence of peanut and tree nut allergy in the United States determined by means of a random digit dial telephone survey: a 5-year follow-up study. *Journal of Allergy and Clinical Immunology*, **112**, 1203–7.

Siegert, R. J. and Ward, T. (2002). Clinical psychology and evolutionary psychology: toward a dialogue. *Review of General Psychology*, **6**, 235–59.

Simhon, A., Douglas, J. R., Drasar, B. S. and Sotthill, J. F. (1982). Effect of feeding on infants' faecal flora. *Archives of Disease in Childhood*, **57**, 54–8.

Simmons, R. (2005). Developmental origins of adult metabolic disease: concepts and controversies. *TRENDS in Endocrinology and Metabolism*, **16**, 390–4.

Sinclair, A. and O'Dea, K. (1993). The significance of arachidonic acid in hunter–gatherer diets: implications for the contemporary western diet. *Journal of Food Lipids*, **1**, 143–57.

Slattery, M. L. and Randall, D. E. (1988). Trends in coronary heart disease mortality and food consumption in the United States between 1909 and 1980. *American Journal of Clinical Nutrition*, **47**, 1060–70.

Smith, B. H. (1991). Dental development and the evolution of life history in Hominidae. *American Journal of Physical Anthropology*, **86**, 157–74.

Smith, C. J. (1994). Food habit and cultural changes among the Pima Indians. In *Diabetes as a Disease of Civilization*, ed. J. R. Joe and R. S. Young. New York: Mouton de Gruyter.

Smith, C. J., Nelson, R. G., Hardy, S. A., Manahan, E. M., Bennett, P. H. and Knowler, W. C. (1996). Survey of the diet of Pima Indians using quantitative food frequency assessment and 24-hour recall. *Journal of the American Dietetic Association*, **96**, 778–84.

Smith, E. O. (2002). *When Culture and Biology Collide: Why we are Stressed, Depressed, and Self-obsessed*. New Brunswick, New Jersey: Rutgers University Press.

Smith, M. T. (1993). Genetic adaptation. In *Human Adaptation*, ed. G. A. Harrison. Oxford: Oxford University Press, pp. 1–54.

Smyth, J., Ockenfels, M. C., Porter, L., Kirschbaum, C., Hellhammer, D. H. and Stone, A. A. (1998). Stressors and mood measured on a momentary basis are associated with salivary cortisol secretion. *Psychoneuroendocrinology*, **23**, 353–70.

Snijder, M. B., van Dam, R. M., Visser, M. and Seidell, J. C. (2005). What aspects of body fat are particularly hazardous and how do we measure them? *International Journal of Epidemiology*, **35**, 83–92.

Solomon, C. G. (1999). The epidemiology of polycystic ovary syndrome. *Endocrinology and Metabolism Clinics of North America*, **28**, 247–63.

Sood, A., Ford, E. S. and Camargo, C. A. (2006). Association between leptin and asthma in adults. *Thorax*, **61**, 300–5.

Sowers, J. R. (2003). Obesity as a cardiovascular risk factor. *American Journal of Medicine*, **115**, 37S–41S.

Spiegel, K., Knutson, K., Leproult, R., Tasali, E. and Van Cauter, E. (2005). Sleep loss: a novel risk factor for insulin resistance and type 2 diabetes. *Journal of Applied Physiology*, **99**, 2008–19.

Spielman, R. S., Fajans, S. S., Neel, J. V., Pek, S., Floyd, J. C. and Oliver, W. J. (1982). Glucose tolerance in two unacculturated Indian tribes of Brazil. *Diabetologia*, **23**, 90–3.

Stamatakis, E., Primatesta, P., Chinn, S., Rona, R. and Falascheti, E. (2005). Overweight and obesity trends from 1974 to 2003 in English children: what is the role of socioeconomic factors? *Archives of Disease in Childhood*, **90**, 999–1004.

Stampfer, M. J. (2006). Cardiovascular disease and Alzheimer's disease: common links. *Journal of Internal Medicine*, **260**, 211–23.

Stampfer, M. J. and Rimm, E. B. (1995). Epidemiologic evidence for vitamin E in prevention of cardiovascular disease. *American Journal of Clinical Nutrition*, **62**, S1365–9.

Stanhope, J. M. and Prior, I. A. M. (1980). The Tokelau island migrant study: prevalence and incidence of diabetes mellitus. *New Zealand Medical Journal*, **92**, 417–21.

Stearns, S. C. and Ebert, D. (2001). Evolution in health and disease: work in progress. *Quarterly Review of Biology*, **76**, 417–32.

Steckel, R. H. and Rose, J. C. (2002b). Conclusions. In *The Backbone of History: Health and Nutrition in the Western Hemisphere*, ed. R. H. Steckel and J. C. Rose. Cambridge: Cambridge University Press, pp. 583–9.

Steckel, R. H. and Rose, J. C. (2002a). Patterns of health in the western hemisphere. In *The Backbone of History: Health and Nutrition in the Western Hemisphere*, ed. R. H. Steckel and J. C. Rose. Cambridge: Cambridge University Press, pp. 563–79.

Steckel, R. H., Rose, J. C., Larsen, C. S. and Walker, P. L. (2002). Skeletal health in the Western Hemisphere from 4000BC to the present. *Evolutionary Anthropology*, **11**, 142–55.

Steffen, P. R., McNeilly, M., Anderson, N. and Sherwood, A. (2003). Effects of perceived racism and anger inhibition on ambulatory blood pressure in African Americans. *Psychosomatic Medicine*, **65**, 746–50.

Stene, L. C. and Nafstad, P. (2001). Relation between occurrence of type 1 diabetes and asthma. *The Lancet*, **357**, 607–8.

Steptoe, A. and Marmot, M. (2002). The role of psychobiological pathways in socio-economic inequalities in cardiovascular disease risk. *European Heart Journal*, **23**, 13–25.

Steptoe, A., Wardle, J., Lipsey, Z., Oliver, G., Pollard, T. M. and Davies, G. J. (1998). The effects of life stress on food choice. In *The Nation's Diet: The Social Science of Food Choice*, ed. A. Murcott. Harlow: Addison Wesley Longman, pp. 29–42.

Stevens, R. G. (2006). Artificial lighting in the industrialized world: circadian disruption and breast cancer. *Cancer Causes and Control*, **17**, 501–7.

Stevens, R. G. and Rea, M. S. (2001). Light in the built environment: potential role of circadian disruption in endocrine disruption and breast cancer. *Cancer Causes and Control*, **12**, 279–87.

Stinson, S. (2002). Early childhood health in foragers. In *Human Diet: Its Origin and Evolution*, ed. P. S. Ungar and M. F. Teaford. Westport, Connecticut: Bergin and Garvey, pp. 37–48.

Storey, R. (1985). An estimate of mortality in a Pre–Columbian urban population. *American Anthropologist*, **83**, 519–35.

Strachan, D. P. (1989). Hay fever, hygiene and household size. *British Medical Journal*, **299**, 1259–60.

Strachan, D. P. (1997). Allergy and family size: a riddle worth solving. *Clinical and Experimental Allergy*, **27**, 235–6.

Strassmann, B. I. (1999). Menstrual cycling and breast cancer: an evolutionary perspective. *Journal of Women's Health*, **8**, 193–202.

Strassmann, B. I. and Dunbar, R. I. M. (1999). Human evolution and disease: putting the Stone Age in perspective. In *Evolution in Health and Disease*, ed. S. C. Stearns. Oxford: Oxford University Press, pp. 91–101.

Strauss, R. S. and Pollack, H. A. (2001). Epidemic increase in childhood overweight, 1986–1998. *Journal of the American Medical Association*, **286**, 2845–8.

Stringer, C. (2002). Modern human origins: progress and prospects. *Philosophical Transactions of the Royal Society of London B*, **357**, 563–79.

Stringer, C. (2003). Out of Ethiopia. *Nature*, **423**, 692–5.

Stuart-Macadam, P. (1995). Breastfeeding in prehistory. In *Breastfeeding: Biocultural Perspectives*, ed. P. Stuart-Macadam and K. A. Dettwyler. New York: Aldine de Gruyter, pp. 75–99.

Stuart-Macadam, P. (1998). Iron deficiency anemia: exploring the difference. In *Sex and Gender in Paleopathological Perspective*, ed. A. L. Grauer and P. Stuart-Macadam. Cambridge: Cambridge University Press, pp. 45–63.

Sutton-Tyrrell, K., Wildman, R. P., Matthews, K. A. *et al.* (2005). Sex hormone-binding globulin and the free androgen index are related to cardiovascular risk factors in multiethnic premenopausal women and perimenopausal women enrolled in the study of women across the nation (SWAN). *Circulation*, **111**, 1242–9.

Swinburn, B. and Egger, G. (2004). The runaway weight gain train: too many accelerators, not enough brakes. *British Medical Journal*, **329**, 736–9.

Szathmáry, E. J. E. (1994). Non-insulin dependent diabetes mellitus among Aboriginal North Americans. *Annual Review of Anthropology*, **23**, 457–82.

Taheri, S. (2006). The link between short sleep duration and obesity: we should recommend more sleep to prevent obesity. *Archives of Disease in Childhood*, **91**, 881–4.

Taheri, S., Lin, L., Austin, D., Young, T. and Mignot, E. (2004). Short sleep duration is associated with reduced leptin, elevated ghrelin, and increased body mass index. *PLoS Medicine*, **1**, 210–17.

Talayero, J. M. P., Lizan-Garcia, M., Puime, A. O. *et al.* (2006). Full breastfeeding and hospitalization as a result of infections in the first year of life. *Pediatrics*, **118**, E92–9.

Tatz, C. (2005). *Aboriginal Suicide is Different: Portrait of Life and Self-Destruction*. Canberra: Aboriginal Studies Press.

Taylor, A. L., Dunstan, J. A. and Prescott, S. L. (2007). Probiotic supplementation for the first 6 months of life fails to reduce the risk of atopic dermatitis and increases the risk of allergen sensitization in high-risk children: a randomized controlled trial. *Journal of Allergy and Clinical Immunology*, **119**, 184–91.

Taylor, J. S., Kacmar, J. E., Nothnagle, M. and Lawrence, R. A. (2005). A systematic review of the literature associating breastfeeding with type 2 diabetes and gestational diabetes. *Journal of the American College of Nutrition*, **24**, 320–6.

Tesfaye, F., Nawi, N. G., Van Minh, H. *et al.* (2007). Association between body mass index and blood pressure across three populations in Africa and Asia. *Journal of Human Hypertension*, **21**, 28–37.

Thakore, J. H., Richards, P., Reznek, R. H., Martin, A. and Dinan, T. G. (1997). Increased intra-abdominal fat deposition in patients with major depressive illness as measured by computed tomography. *Biological Psychiatry*, **41**, 1140–2.

Thomas, R. B. (1998). The evolution of human adaptability paradigms: toward a biology of poverty. In *Building a New Biocultural Synthesis*, ed. A. H. Goodman and T. L. Leatherman. Ann Arbor: University of Michigan Press, pp. 43–74.

Thomson, N. J. (1991). Recent trends in aboriginal mortality. *Medical Journal of Australia*, **154**, 235–9.

Thorburn, A. W. (2005). National prevalence of obesity – prevalence of obesity in Australia. *Obesity Reviews*, **6**, 187–9.

Tillin, T., Forouhi, N., Johnston, D. G., McKeigue, P., Chaturvedi, N. and Godsland, I. F. (2005). Metabolic syndrome and coronary heart disease in South Asians, African-Caribbeans and white Europeans: a UK population-based cross-sectional study. *Diabetologia*, **48**, 649–56.

Tishkoff, S. A. and Kidd, K. K. (2004). Implications of biogeography of human populations for 'race' and medicine. *Nature Genetics*, **36**, S21–7.

Tishkoff, S. A., Reed, F. A., Ranciaro, A. *et al.* (2007). Convergent adaptation of human lactase persistence in Europe. *Nature Genetics*, **39**, 31–40.

Tominaga, S. and Kuroishi, T. (1997). An ecological study on diet/nutrition and cancer in Japan. *International Journal of Cancer*, **71**, S10, 2–6.

Tonetti, D. (2004). Prevention of breast cancer by recapitulation of pregnancy hormone levels. *Breast Cancer Research*, **6**, E8.

Travis, R. C., Allen, D. S., Fentiman, I. S. and Key, T. J. (2004). Melatonin and breast cancer: a prospective study. *Journal of the National Cancer Institute*, **96**, 475–82.

Treloar, A. E., Boynton, R. E., Behn, B. G. and Brown, B. W. (1967). Variation of the human menstrual cycle through reproductive life. *International Journal of Fertility*, **12**, 77–126.

Tremblay, M. S., Katzmarzyk, P. T. and Willms, J. D. (2002). Temporal trends in overweight and obesity in Canada 1981–1996. *International Journal of Obesity*, **26**, 538–43.

Trevathan, W. R., Smith, E. O. and McKenna, J. J., eds. (1999). *Evolutionary Medicine*. New York: Oxford University Press.

Trevathan, W. R., Smith, E. O. and McKenna, J. J., eds. (2007). *New Perspectives in Evolutionary Medicine*. New York: Oxford University Press.

Trichopoulos, D., MacMahon, B. and Cole, P. (1972). Menopause and breast cancer risk. *Journal of the National Cancer Institute*, **48**, 605–13.

Triggs-Raine, B. L., Kirkpatrick, R. D., Kelly, S. L. *et al.* (2002). HNF-1alpha G319S, a transactivation-deficient mutant, is associated with altered dynamics of diabetes onset in an Oji-Cree community. *Proceedings of the National Academy of Science*, **99**, 4614–19.

Troiano, R. P., Briefel, R. R., Carroll, M. D. and Bialostosky, K. (2000). Energy and fat intakes of children and adolescents in the United States: data from the National Health and Nutrition Examination Surveys. *American Journal of Clinical Nutrition*, **72**, 1343S–53S.

Trowell, H. C. and Burkitt, D. P. (1981a). Preface. In *Western Diseases: Their Emergence and Prevention*, ed. H. C. Trowell and D. P. Burkitt. Cambridge, Massachusetts: Harvard University Press, pp. xiii–xvi.

Trowell, H. C. and Burkitt, D. P., eds. (1981b). *Western Diseases: Their Emergence and Prevention*. Cambridge, Massachusetts: Harvard University Press.

Truswell, A. S. and Hansen, J. D. L. (1976). Medical research among the !Kung. In *Kalahari Hunter–Gatherers: Studies of the !Kung San and their Neighbours*, ed. R. B. Lee and I. DeVore. Cambridge, Massachusetts: Harvard University Press, pp. 166–94.

Tunstall-Pedoe, H., Connaghan, J., Woodward, M., Tolonen, H. and Kuulasmaa, K. (2006). Pattern of declining blood pressure across replicate population surveys of the WHO MONICA project, mid-1980s to mid-1990s, and the role of medication. *British Medical Journal*, **332**, 629–35.

Ueshima, H., Okayama, A., Saitoh, S. *et al.* (2003). Differences in cardiovascular disease risk factors between Japanese in Japan and Japanese-Americans in Hawaii: the INTERLIPID study. *Journal of Human Hypertension*, **17**, 631–9.

Umetsu, D. T. and DeKruyff, R. H. (2006). The regulation of allergy and asthma. *Immunological Reviews*, **212**, 238–55.

Uusitalo, U., Feskens, E. J. M., Tuomilehto, J. *et al.* (1996). Fall in total cholesterol concentration over five years in association with changes in fatty acid composition of cooking oil in Mauritius: cross sectional survey. *British Medical Journal*, **313**, 1044–6.

van Anders, S. M. and Watson, N. V. (2006). Menstrual cycle irregularities are associated with testosterone levels in healthy premenopausal women. *American Journal of Human Biology*, **18**, 841–4.

Van Blerkom, L. (2003). Role of viruses in human evolution. *Yearbook of Physical Anthropology*, **46**, 14–46.

van der Klink, J. J. L., Blonk, R. W. B., Schene, A. H. and van Dijk, F. J. H. (2001). The benefits of interventions for work-related stress. *American Journal of Public Health*, **91**, 271–6.

van der Spuy, Z. M. and Dyer, S. J. (2004). The pathogenesis of infertility and early pregnancy loss in polycystic ovary syndrome. *Best Practice and Research Clinical Obstetrics and Gynaecology*, **18**, 755–71.

van Eck, M. M. and Nicolson, N. A. (1994). Perceived stress and salivary cortisol in daily life. *Annals of Behavioral Medicine*, **16**, 221–7.

van Hooff, M. H. A., Voorhorst, F. J., Kaptein, M. B. H., Hirasing, R. A., Koppenaal, C. and Schoemaker, J. (2004). Predictive value of menstrual cycle pattern, body mass index, hormone levels and polycystic ovaries at age 15 years for oligo-amenorrhoea at age 18 years. *Human Reproduction*, **19**, 383–92.

van Odijk, J., Kull, I., Borres, M. P. *et al.* (2003). Breastfeeding and allergic disease: a multidisciplinary review of the literature (1966–2001) on the mode of early feeding in infancy and its impact on later atopic manifestations. *Allergy*, **58**, 833–43.

Van Poppel, G., Kardinaal, A., Princen, H. and Kok, F. J. (1994). Antioxidants and coronary heart disease. *Annals of Medicine*, **26**, 429–34.

Vartiainen, E., Jousilahti, P., Alfthan, G., Sundvall, J., Pietinen, P. and Puska, P. (2000). Cardiovascular risk factors in Finland, 1972–1997. *International Journal of Epidemiology*, **29**, 49–56.

Verkasalo, P. K., Thomas, H. V., Appleby, P. N., Davey, G. K. and Key, T. J. (2001). Circulating levels of sex hormones and their relation to risk factors for breast cancer: a cross-sectional study in 1092 pre- and post-menopausal women (United Kingdom). *Cancer Causes and Control*, **12**, 47–59.

Villamor, E. and Cnattingius, S. (2006). Interpregnancy weight change and risk of adverse pregnancy outcomes: a population-based study. *The Lancet*, **368**, 1164–70.

Virtanen, S. M. and Knip, M. (2003). Nutritional risk predictors of β cell autoimmunity and type 1 diabetes at a young age. *American Journal of Clinical Nutrition*, **78**, 1053–67.

Vitzthum, V. J. (2001). Why not so great is still good enough. In *Reproductive Ecology and Human Evolution*, ed. P. T. Ellison. New York: Aldine de Gruyter, pp. 179–202.

Vitzthum, V. J., Bentley, G. R., Spielvogel, H. *et al.* (2002). Salivary progesterone levels and rate of ovulation are significantly lower in poorer than in better-off urban-dwelling Bolivian women. *Human Reproduction*, **17**, 1906–13.

Vitzthum, V. J., Spielvogel, H. and Thornburg, J. (2004). Interpopulational differences in progesterone levels during conception and implantation in humans. *Proceedings of the National Academy of Science*, **101**, 1443–8.

Vogel, V. G., Costantino, J. P., Wickerham, D. L. *et al.* for the National Surgical Adjuvant Breast and Bowel Project (NSABP) (2006). Effects of tamoxifen vs raloxifene on the risk of developing invasive breast cancer and other disease outcomes: the NSABP study of tamoxifen and raloxifene (STAR) P-2 trial. *Journal of the American Medical Association*, **295**, 2727–41.

von Ehrenstein, O. S., von Mutius, E., Illi, S., Baumann, L., Bohm, O. and von Kries, R. (2000). Reduced risk of hay fever and asthma among children of farmers. *Clinical and Experimental Allergy*, **30**, 187–93.

von Hertzen, L. and Haahtela, T. (2005). Signs of reversing trends in prevalence of asthma. *Allergy*, **60**, 283–92.

von Mutius, E., Martinez, F. D., Fritzsch, C., Nicolai, T., Reitmeir, P. and Thiemann, H.-H. (1994). Skin test reactivity and number of siblings. *British Medical Journal*, **308**, 692–5.

Vorona, R. D., Winn, M. P., Babineau, T. W., Eng, B. P., Feldman, H. R. and Ware, J. C. (2005). Overweight and obese patients in a primary care population report less sleep than patients with a normal body mass index. *Archives of Internal Medicine*, **165**, 25–30.

Waldron, I. (1991). Patterns and causes of gender differences in smoking. *Social Science and Medicine*, **32**, 989–1005.

Wallace, B. A. and Cumming, R. G. (2000). Systematic review of randomized trials of the effect of exercise on bone mass in pre- and postmenopausal women. *Calcified Tissue International*, **67**, 10–18.

Walters, V. (1993). Stress, anxiety and depression: women's accounts of their health problems. *Social Science and Medicine*, **36**, 393–402.

Wang, C., Catlin, D. H., Starcevic, B. *et al.* (2005). Low-fat high-fiber diet decreased serum and urine androgens in men. *Journal of Clinical Endocrinology and Metabolism*, **90**, 3550–9.

Wang, Y. Z., Mi, J., Shan, X.-Y., Wang, Q. J. and Ge, K.-Y. (2007). Is China facing an obesity epidemic and the consequences? The trends in obesity and chronic disease in China. *International Journal of Obesity*, **31**, 177–88.

Warren, M. (2004). A comparative review of the risks and benefits of hormone replacement therapy regimens. *American Journal of Obstetrics and Gynecology*, **190**, 1141–67.

Watts, J. (2006). Doctors blame air pollution for China's asthma increases. *The Lancet*, **368**, 719–720.

Weedon, M. N., Schwarz, P. E. H., Horikawa, Y. *et al.* (2003). Meta-analysis and a large association study confirm a role for Calpain-10 variation in type 2 diabetes susceptibility. *American Journal of Human Genetics*, **73**, 1208–12.

Weir, G., Laybutt, D., Kaneto, H., Bonner-Weir, S. and Sharma, A. (2001). Beta-cell adaptation and decompensation during the progression to diabetes. *Diabetes*, **50** (**Suppl.1**), 5154–9.

Weiss, K. M. and Fullerton, S. M. (2005). Racing around, getting nowhere. *Evolutionary Anthropology*, **14**, 165–9.

Weiss, K. M., Ferrell, R. E. and Hanis, C. L. (1984). A New World Syndrome of metabolic diseases with a genetic and evolutionary basis. *Yearbook of Physical Anthropology*, **27**, 153–78.

Wells, J. C. K. (2006). The evolution of human fatness and susceptibility to obesity: an ethological approach. *Biological Reviews*, **81**, 183–205.

Wenzel, S. E. (2006). Asthma: defining of the persistent adult phenotypes. *The Lancet*, **368**, 804–813.

West, K. M. (1974). Diabetes in American Indians and other native populations of the New World. *Diabetes*, **23**, 841–55.

West, R. (2006). Tobacco control: present and future. *British Medical Bulletin*, **77–78**, 123–36.

Whitmer, R. A., Gunderson, E. P., Barrett-Connor, E., Quesenberry, C. P. and Yaffe, K. (2005). Obesity in middle age and future risk of dementia: a 27 year longitudinal population based study. *British Medical Journal*, **330**, 1360–2.

WHO Expert Consultation (2004). Appropriate body-mass index for Asian populations and its implications for policy and intervention strategies. *The Lancet*, **363**, 157–63.

WHO Global Infobase Team (2005). *The SuRF Report 2. Surveillance of Chronic Disease Risk Factors: Country-Level Data and Comparable Estimates*. Geneva: World Health Organization.

WHO International Consortium in Psychiatric Epidemiology (2000). Cross-national comparisons of the prevalences and correlates of mental disorders. *Bulletin of the World Health Organization*, **78**, 413–26.

Wild, S., Roglic, G., Green, A., Sicree, R. and King, H. (2004). Global prevalence of diabetes: estimates for the year 2000 and projections for 2030. *Diabetes Care*, **27**, 1047–53.

Wild, S. H. and Byrne, C. D. (2006). Risk factors for diabetes and coronary heart disease. *British Medical Journal*, **333**, 1009–11.

Wilkinson, R. (1999). Health, hierarchy, and social anxiety. *Annals of the New York Academy of Sciences*, **896**, 48–63.

Williams, B. (1995). Westernised Asians and cardiovascular disease: nature or nurture. *The Lancet*, **345**, 401–2.

Williams, D. R. and Collins, C. (1995). US socioeconomic and racial differentials in health: patterns and explanations. *Annual Review of Sociology*, **21**, 349–86.

Williams, D. R., Neighbors, H. W. and Jackson, J. S. (2003). Racial/ethnic discrimination and health: findings from community studies. *American Journal of Public Health*, **93**, 200–8.

Williams, G. C. (1957). Pleiotropy, natural selection, and the evolution of senescence. *Evolution*, **11**, 398–411.

Williams, G. C. and Nesse, R. M. (1991). The dawn of Darwinian medicine. *Quarterly Review of Biology*, **66**, 1–22.

Williams, R. C., Long, J. C., Hanson, R. L., Sievers, M. L. and Knowler, W. C. (2000). Individual estimates of European genetic admixture associated with lower body-mass index, plasma glucose, and prevalence of type 2 diabetes in Pima Indians. *American Journal of Human Genetics*, **66**, 527–38.

Wilmoth, J. R. (2000). Demography of longevity: past, present, and future trends. *Experimental Gerontology*, **35**, 1111–29.

Wilson, B. D., Wilson, N. C. and Russell, D. G. (2001). Obesity and body fat distribution in the New Zealand population. *New Zealand Medical Journal*, **114**, 127–30.

Wilson, T. W. and Grim, C. E. (1991). Biohistory of slavery and blood pressure differences in blacks today. A hypothesis. *Hypertension*, **17** (Suppl. 1), 1122–9.

Wood, B. and Collard, M. (1999). The human genus. *Science*, **284**, 65–71.

World Health Organization (1999). *World Health Report: Making a Difference.* Geneva, World Health Organization.

World Health Organization (2003). *Diet, Nutrition and the Prevention of Chronic Diseases.* Geneva: World Health Organization.

Worthman, C. M. and Melby, M. K. (2002). Toward a comparative developmental ecology of human sleep. In *Adolescent Sleep Patterns: Biological, Social and Psychological Influences*, ed. M. A. Carskadon. New York: Cambridge University Press, pp. 69–117.

Wrigley, E. A. (1969). *Population and History.* London: Weidenfeld and Nicolson.

Writing Group for the Women's Health Initiative Investigators (2002). Risks and benefits of estrogen plus progestin in healthy postmenopausal women: principal results from the Women's Health Initiative randomized controlled trial. *Journal of the American Medical Association*, **288**, 321–33.

Wu, A. H., Pike, M. C. and Stram, D. O. (1999). Meta-analysis: dietary fat intake, serum estrogen levels, and the risk of breast cancer. *Journal of the National Cancer Institute*, **91**, 529–34.

Wu, Y. (2006). Overweight and obesity in China. *British Medical Journal*, **333**, 362–3.

Yajnik, C. (2000). Interactions of perturbations in intrauterine growth and growth during childhood on the risk of adult-onset disease. *Proceedings of the Nutrition Society*, **59**, 257–65.

Yajnik, C., Lubree, H., Rege, S. *et al.* (2002). Adiposity and hyperinsulinemia in Indians are present at birth. *Journal of Clinical Endocrinology and Metabolism*, **87**, 5575–80.

Yang, L., Parkin, D. M., Ferlay, J., Li, L. and Chen, Y. (2005). Estimates of cancer incidence in China for 2000 and projections for 2005. *Cancer Epidemiology, Biomarkers and Prevention*, **14**, 243–50.

Yazdanbakhsh, M., Kremsner, P. G. and van Ree, R. (2002). Allergy, parasites, and the hygiene hypothesis. *Science*, **296**, 490–4.

Yemaneberhan, H., Bekele, Z., Venn, A., Lewis, S., Parry, E. and Britton, J. (1997). Prevalence of wheeze and asthma and relation to atopy in urban and rural Ethiopia. *The Lancet*, **350**, 85–90.

Yemaneberhan, H., Flohr, C., Lewis, S. A. *et al.* (2004). Prevalence and associated factors of atopic dermatitis symptoms in rural and urban Ethiopia. *Clinical and Experimental Allergy*, **34**, 779–85.

Yoon, K.-H., Lee, J.-H., Kim, J.-W. *et al.* (2006). Epidemic obesity and type 2 diabetes in Asia. *The Lancet*, **368**, 1681–8.

Young, D. B., Lin, H. and McCabe, R. D. (1995). Potassium's cardiovascular protective mechanisms. *American Journal of Physiology: Regulatory, Integrative and Comparative Physiology*, **37**, R825–37.

Young, J. H., Chang, Y. P. C., Kim, J. D. O. *et al.* (2005). Differential susceptibility to hypertension is due to selection during the out-of-Africa expansion. *PLoS Genetics*, **1**, 730–8.

Young, T. K., Reading, J., Elias, B. and O'Neil, J. D. (2000). Type 2 diabetes mellitus in Canada's First Nations: status of an epidemic in progress. *Canadian Medical Association Journal*, **163**, 561–6.

Yusuf, S., Reddy, S., Ounpuu, S. and Anand, S. (2001a). Global burden of cardiovascular diseases. Part I: General considerations, the epidemiologic transition, risk factors, and the impact of urbanization. *Circulation*, **104**, 2746–53.

Yusuf, S., Reddy, S., Ôunpuu, S. and Anand, S. (2001b). Global burden of cardiovascular diseases. Part II: Variations in cardiovascular disease by specific ethnic groups and geographic regions and prevention strategies. *Circulation*, **104**, 2855–64.

Zainudin, B. M. Z., Lai, C. K. W., Sporiano, J. B., Jia-Horng, W. and De Guia, T. S. (2005). Asthma control in adults in Asia–Pacific. *Respirology*, **10**, 579–86.

Zhou, B. F., Stamler, J., Dennis, B. *et al.* (2003). Nutrient intakes of middle-aged men and women in China, Japan, United Kingdom, and United States in the late 1990s: the INTERMAP study. *Journal of Human Hypertension*, **17**, 623–30.

Ziegler, R. G., Hoover, R. N., Pike, M. C. *et al.* (1993). Migration patterns and breast cancer risk in Asian-American women. *Journal of the National Cancer Institute*, **85**, 1819–27.

Zimmet, P. (2000). Globalization, coca-colonization and the chronic disease epidemic: can the Doomsday scenario be averted? *Journal of Internal Medicine*, **247**, 301–10.

Zimmet, P., Taft, P., Guinea, A., Guthrie, W. and Thoma, K. (1977). The high prevalence of diabetes mellitus on a central Pacific island. *Diabetologia*, **13**, 111–15.

Zimmet, P., Faaiuso, S., Ainuu, J., Whitehouse, S., Milne, B. and DeBoer, W. (1981). The prevalence of diabetes in the rural and urban Polynesian population of Western Samoa. *Diabetes*, **30**, 45–51.

Zizza, C., Siega-Riz, A. M. and Popkin, B. M. (2001). Significant increases in young adults' snacking between 1977–1978 and 1994–1996 represents a cause of concern! *Preventive Medicine*, **32**, 303–10.

Zografos, G. C., Panou, M. and Panou, N. (2004). Common risk factors of breast and ovarian cancer: recent view. *International Journal of Gynecogical Cancer*, **14**, 721–40.

Zoratti, R. (1998). A review on ethnic differences in plasma triglyceride and high-density-lipoprotein cholesterol: is the lipid pattern the key factor for the low coronary heart disease rate in people of African origin? *European Journal of Epidemiology*, **14**, 9–21.

Index

abdominal fat
   in Asian populations 160
   in babies 67
   and breast cancer 83
   and cortisol 148
   historical trends in United States 46
   and risk of type 2 diabetes and
      cardiovascular disease 35
Ache
   mortality 12, 111
   physical activity in 28
   testosterone levels in 94, 95
adaptation
   developmental 5, 154
   phenotypic 5
adipose tissue
   in hominin evolution 30–1
   and inflammation 43
   and oestrogen in postemenopausal women
      92
   patterning on the body 31, 35
   role in storing energy 30
   subcutaneous fat 31
   visceral fat 35 see also
      abdominal fat
adrenaline 144–5, 145–7
   and cardiovascular disease 147
   levels in Western Samoan men 146
age at first birth
   late in western women 97, 141 see also
      breast cancer
ageing population
   global 156
   in western countries 22, 156, 157
agriculture
   early agriculture and health 13–16
air pollution
   and allergies 124–5
   ozone 124
   smog 17
   sulphur dioxide 124
alcohol
   human preference for 28
   and negative emotions 149
allergens 123

allergy
   to foods 121, 123
   genetic vulnerability to 132–4
   prevention of 170 see also
      asthma; hay fever; eczema
allostatic load 148
Alzheimer's disease 48
   in postmenopausal women 113
amenorrhoea 100
androgens
   and impaired reproductive function in
      women 100–2
   in men 82–3
   and prostate cancer 82
andropause, see menopause, male
antidepressants 137
antioxidants 27
   and cardiovascular disease 44
appetite control systems 38
   and breastfeeding 107
   and sleep 42
Ascaris, see helminth infections
asthma 120–1
   and air pollution 124
   allergic 120
   and breastfeeding 130
   non-allergic 120
   and obesity 131–2
   and poverty 133
   prevalence worldwide 122, 123
atherosclerosis 44
   decline over twentieth century 45
   and hypertension 44
   in post-menopausal women 113
   and stress 148
atopy 121
Australian Aborigines
   diet in traditional hunter–gatherer groups
      24–5
   effects on health of living as hunter–
      gatherers 163, 164
   obesity and type 2 diabetes in 53
   physical activity in traditional hunter–
      gatherer groups 28
   socio-economic disadvantage 73

Australian Aborigines *(cont.)*
    thrifty phenotype in 69
autoimmunity 108, 170

Barker, David 5, 66, 69, 72
bilharzia, *see* schistosomiasis
birthweight
    and abdominal obesity 66
    and blood pressure 66
    and cardiovascular disease 66
    and serum lipids 66
    and type 2 diabetes 66
blood pressure
    and breastfeeding 107
    decline in the west 47
    in post-menopausal woman 113
    projected trends in Africa 162
    in subsistence level societies 31
    see also hypertension
body mass index (BMI)
    definition of 31
    values used to define overweight and
        obesity 32
Boyden, Steven 1, 9, 16, 28
brain size
    and age at weaning 104
    energetic costs and hominin evolution 3, 31
    relationship to adiposity 30
breast cancer 76
    and changes in light exposure 98
    in China 161
    and disruption of circadian rhythm 85
    genes 78
    and hormone replacement therapy 84, 114
    incidence worldwide 77
    and late first birth or nulliparity 84, 98
    and melatonin 85, 98
    preventative drugs 169
breastfeeding 103, 105–6, 106–10
    and allergy 130–1
    benefits of 170
    and cognitive development 108, 109
    and disorders of immune regulation 107
    energetic costs 30
    and infectious disease 106
    and insulin resistance in the mother 106
    wet nurses 104
    and women's employment 105
Burkitt, Denis 1

Caesarian delivery 106
    and gut flora colonisation 127
cancer 18, 75
    in China 161
    colon 48

endometrial 76
gallbladder 48
genes 78
kidney 48
liver 48
lung 7, 22, 76, 161
oesophageal 48
ovarian 76, 77
pancreatic 48
testicular 77 *see also*
        breast cancer; prostate cancer
cardiovascular disease 19, 43–8
    in Asia 159
    and hormone replacement therapy 114
    increases associated with urbanisation in
        poorer countries 48
    populations with unusually high rates of 50–5
    prevention of 166–7
    trends in mortality and morbidity in the
        west 45–8 *see also*
        heart disease; hypertension; stroke
cars 41
    efforts to reduce reliance on 166
    increased ownership in Asia 159
    and stress 142
cereal consumption
    in early agriculturalists 14
    refined 39
Chlamydia pneumoniae 45
cholesterol level, *see* serum lipid profile
climacteric 117
clinical depression 136, 137, 140
    projected global burden 156
coeliac disease 107
Cohen, Mark Nathan 9, 10, 12, 13, 26, 31
control, lack of 142, 150
convenience foods 38
coronary heart disease
    decline in the twentieth century 22, 47
    high rates in African-Americans 72
    low rates in British Afro-Caribbean men 55
    projected global burden 156
    psychosocial risk factors 143
    and socio-economic status 47, 72
    underlying pathology 43–4
cortisol 95, 144, 145
    disruption of circadian rhythm 91
    effects of high levels 148
    low levels 148
    variation in everyday life 145, 147
co-sleeping, mother–infant 108
cot death, *see* Sudden Infant Death Syndrome
        (SIDS)
Crohn's disease 129
    and *Trichuris* infection 130

cultural consensus 151
cytokines 35
    inflammatory 43
    anti-inflammatory 129

Darwinian medicine 2
dementia 48
dental health 15
depression
    adaptive explanations for 139
    and health-related behaviour 149
    prevention of 171
developmental effects on ovarian function 89–90
developmental origins of adult disease 66–8
    in rapidly westernising populations 68–70
    role of maternal nutrition 69
    role of maternal smoking 69
developmental plasticity, *see* adaptation, developmental
diabesity 32
diabetes, *see* gestational diabetes; maturity-onset diabetes-of-the-young; type 1 diabetes; type 2 diabetes
dietary fibre
    decline in the west 39
    in hunter–gatherer diet 27
disability-adjusted life years (DALYs), definition of 34
diseases of affluence 1–2
divorce 141
dust mites 123, 163
dyslipidaemia 39
eating disorders 165
eczema 121, 122
    and breastfeeding 130
egalitarianism 10
    in the Kaluli 142
    and social anxiety 140

Ellison, Peter 75, 88, 89, 94, 96, 118, 169
email 141
energy expenditure
    decline in Asia 159
    in hunter–gatherers 26
    in thermoregulation 30 *see also* physical activity
energy intake
    trends over the twentieth century 38
Enterobius, *see* helminth infections
environment of evolutionary adaptedness 4
epidemiological transition 6, 18–21, 156
epinephrine, *see* adrenaline
evolutionary medicine 2
evolutionary psychology 139, 140

famine, decline of 18
fast food 38
fat consumption
    and appetite control 38
    decline in the west 40, 47
    in hunter–gatherers 26, 27
    increases in China and elsewhere 48, 160
    and ovarian hormone levels 91, 168
    and serum lipid levels 44
    and testosterone levels 97
fat-free mass 34
fight or flight response 140, 144
food security
    in early agriculturalists 14
    in hunter–gatherers 27
    and the non-thrifty genotype 60
    and the thrifty genotype 55, 56
foragers, *see* hunter–gatherers
fracture 113
    and hormone replacement therapy 114

genetic resistance to infectious disease 15
gestational diabetes 101, 106
ghrelin 42
    and immunoglobulin E 132
glucose, serum 36, 38
glycaemic index 38–9
grandmother hypothesis 111

Hadza
    body mass index 31
    diet 26
    role of grandmothers 111
Harrison, Geoffrey Ainsworth 2, 143
hay fever 121
    and air pollution 124
    and household size 125
heart disease 18
    rheumatic 46, 47 *see also* coronary heart disease
Helicobacter pylori 45
helminth infections
    Ascaris 13
    in early agriculturalists 13
    Enterobius 11, 128
    and the evolution of an inflammatory immune response 133
    in hunter–gatherers 11
    Trichuris 13, 130
HIV/AIDS 21
Homo sapiens
    origins of 3
    migration out of Africa 3

honey
  consumption by hunter–gatherers 24, 27
  and selection against insulin resistance 61
hormone replacement therapy (HRT) 114–15
hot flushes 111, 114, 115, 116
  in men 117
hunter–gatherers
  atherosclerosis in 31
  body composition of 31
  breastfeeding in 104
  diet of 23–31
  ecology and health of 9–13
  energy expenditure of 28–30
  insulin sensitivity in 31
  mortality of 12
  testosterone levels in 94
hygiene hypothesis 125–30, 170
  and autoimmune diseases 129
  childhood infections 125
  gut microflora 127–8
  helminths 126, 128–9
  non-pathogenic bacteria 126
hyperandrogenism in women 100, 101
hyperinsulinaemia 36
  and cancers 48
  and erectile dysfunction 102
hypertension
  and depression 143
  and heat-adapted populations 64, 66
  medical control of 47
  in people of African origin 55, 162
  and stress 148
  underlying pathology 44

immunoglobulin E (IgE) 121
infectious disease
  and allergy
  and cardiovascular disease 45, 46, 71
  decline during epidemiological transition 18
  and inflammation 43
  re-emerging 21
  and type 2 diabetes 71
infertility, female *see* reproductive function
infertility, male 102
inflammation 155
  and allergy 131
  and cardiovascular disease 45, 46
  and obesity and type 2 diabetes 43
inflammatory bowel disease 107, 129
insulin 36, 38
  and impaired reproductive function 99
  and ovarian function 90
insulin gene 61
insulin-like growth factor-1 (IGF-1)
  and cancer 83

and ovarian function 90
insulin resistance 36
  and cancers 48, 83
  and impaired reproductive function 99
  and inflammation 43
  in post-menopausal women 113
intergenerational phenotypic inertia 70
intergenerational transmission of disease risk
      69–70
  epigenetic mechanisms 71
Inuit
  diet and serum lipid profiles 27
  physical fitness 29
iron-deficiency anaemia 14–15
ischaemic heart disease, *see* coronary heart
      disease

job strain 142
job stress 146–7
  strategies to reduce 171

Kleinman, Arthur 137
!Kung
  adiposity 31
  blood pressure 31
  breastfeeding in 104
  diet 26
  mortality 12
  physical activity in 28
  serum lipid level 31

lactational amenorrhoea 105–10
  in the Dogon 94
  in hunter–gatherers 93
lactose, digestion of 15
leptin 42
  and asthma 131
  and ovarian function 90
leukaemia, childhood 107
life expectancy,
  in cities 17, 18
  during the epidemiologic transition 18,
      19
  in early agriculturalists 16
  in hunter–gatherers 12
  in Russia late twentieth century 21
life history theory 89
lifespan perspective 5
long-chain polyunsaturated fatty acids
      (LC-PUFAs) 108

malaria 13, 15
manual work
  decline of 21, 41, 46, 159
  health benefits in Samoans 165
  and the industrial revolution 32

materialism 141
maturity-onset diabetes-of-the-young
    (MODY) 62
meat
    in early agriculturalist diet 13
    in early *Homo sapiens* diet 23
    in hominin evolution 3, 23–4
    human preference for 28
    in hunter–gatherer diet 26
menarche, age at 89, 90
    in western girls 93
menopause 110–17
    age at 93
    definition 110
    evolution of 110–11
    inducing early 169
    male 117
    medicalisation of 113
    'symptoms' of 111, 112
menstrual cycle 80
metabolic syndrome
    definition of 27, 37
milk consumption 15
    and selection against insulin resistance 60–1
multiple sclerosis 107, 129
    and *Trichuris* infection 130

Native Americans
    abdominal obesity in 52
    coronary heart disease in 52
    New World Syndrome 58
    socio-economic disadvantage 73
    thrifty phenotype in 69
    type 2 diabetes in 52
Nauruans
    high rates of type 2 diabetes in 52
    insulin gene in 61
    thrifty genotype in 58
    thrifty phenotype in 69
Neel, James 55, 65, 150
Neolithic revolution 13
Nesse, Randolph 2, 4, 139, 141
non-communicable disease, 1, 18
    global mortality from 19
non-thrifty genotype 60, 70
nulliparity, *see* breast cancer
nutrition transition 48, 160

obesity 23, 32–5
    in Asia 159–61
    and breastfeeding 107
    and cancers 48, 83
    childhood 33, 34, 35
    and hypertension 45
    and impaired reproductive function 99–102

    in middle-income countries 159
    and ovarian hormone levels 90–1
    in Polynesians 51
    prevalence worldwide 33
    prevention 164–6
    projected trends 156, 158
    and testosterone levels 97
    and type 2 diabetes 35
    and serum lipid levels 44 *see also*
        abdominal fat; adipose tissue
obesogenic environment 154, 166
oestradiol
    in post-menopausal women 98
    role in the body 79–82
oestriol 79
oestrogen
    and cancers of the breast, endometrium and
        ovaries 81
    exogenous oestrogens and reproductive
        cancers 84
    and fat deposition 31
    forms of 79
    levels in western women 87
    and osteoporosis 168
    suppression using aromatase inhibitors 169
oestrone 79
    in post-menopausal women 80, 92
Oji-Cree
    genetic vulnerability to type 2 diabetes 63
    ways of life 53, 54
oligomenorrhoea 100
Omram, Abdel 16, 18, 19
organochlorines and disorders of the male
    reproductive tract 84
osteoporosis 113, 116
ovarian function *see* oestrogen and progesterone

pace of life 142
Palaeolithic
    definition 4
    diet, physical activity and body composition
        during 23–31
    human ecology and health during 9–13
pancreas 36, 108
pesticides *see* organochlorines
physical activity
    and abdominal obesity 41
    and breast cancer prevention 168
    and cardiovascular disease 44
    lack of in children 41
    and negative emotions 149
    and osteoporosis 116, 168
    and sleep duration 42
    and treatment of depression 169
    in the west 40–2

physical activity level (PAL) 29
physical fitness
    in hunter–gatherers 29
    and metabolic syndrome 41
phyto-oestrogens
    and free testosterone levels 97
    and oestrogen metabolism 91
    and reproductive cancers 84, 168
Pima
    diet 69
    effect of diabetes during pregnancy 70
    genetic vulnerability to type 2 diabetes 61–2
    menstrual irregularity in 101
    New World Syndrome 58
    type 2 diabetes in 52, 107
plague 17
pinworm, *see* helminths, *Enterobius*
polycystic ovary syndrome (PCOS) 101–2
portion size 38
postpartum depression 139
potassium, dietary 27
    and hypertension 45
poverty, *see* socio-economic status
pregnancy
    energetic costs 30
    weight gain 106
probiotics 170
progesterone
    and cancers of the breast, endometrium and
        ovaries 81
    and energetics 88–91
    levels in western women 86
    role in the body 79–82
    variation in levels across populations 86
prostate cancer 76–7, 82
    and exposure to oestrogen in early life 83
    and foetal exposure to maternal androgens
        96
    incidence worldwide 77
protein consumption in hunter–gatherers 26

race and genetics 65–6, 155
racism
    and blood pressure 72
    and depression 152
    and heart disease 152
reproductive function, impaired 99–103
    in populations in transition 101
roundworm, *see* helminths, *Ascaris*

salivary steroids, assessment of 85
salt consumption
    and blood pressure 45
    efforts to reduce 167
schistosomiasis 13

seasonality
    hunter–gatherer diet 27
    as a selective pressure during hominin
        evolution 30
serum lipid profile
    and atherosclerosis 44
    and breastfeeding 107
    and dietary fat in hunter–gatherer diet 27
    and oestrogen 114
    recent improvements in some populations
        40, 167
sex hormone binding-globulin (SHBG)
    binding oestradiol 80, 86
    binding testosterone in men 82, 83
    and impaired reproductive function 100
    and obesity 91, 92
    and phyto-oestrogens 92
sinking heart 138
slavery hypertension hypothesis 63–4
sleep duration
    and appetite 42
    and increased dietary consumption 42
    and obesity 42–3
    trends 42
    and type 2 diabetes 42
sleep patterns 43
smoking
    and asthma 124, 133
    and cardiovascular disease 45
    decline in the west 47, 167
    and fertility 102
    increases in poorer countries 48, 161
    and lung cancer 7, 22
    and negative emotions 149
    rates in women 103
    and socio-economic status 73
snacking
    by Australian Aborigines 24, 39
    in the United States 39–40
social anxiety 140, 142
social isolation 141
social support 142, 149, 162
socio-economic status
    and breast cancer 87–98
    and depression 141
    and ethnicity 72
    and health 73
    and smoking 151
    and stress 150
    and type 2 diabetes and cardiovascular
        disease 72–3
soft drinks 40
South Asian migrants to the west
    androgen levels in 101
    body composition in 54

genetic vulnerability to type 2 diabetes 59,
    62
heart disease in 54
type 2 diabetes in 54
sperm count 103
Strassmann, Beverly 94, 141
stress
    and cardiovascular disease 147
    definition 136, 137–8
    and health-related behaviour 149
    and immune function 149
stroke
    declining mortality from stroke in the west
        47
    mortality in England and Wales 47
    projected global burden 156
    underlying pathology 43, 44
Sudden Infant Death Syndrome (SIDS) 109
suburbs 166
systemic iupus erythematosus 129

tamoxifen 169
taste preferences 28
T cells
    regulatory T cell 129–30, 171
    T helper 1 (Th1) cells 125, 126, 130
    T helper 2 (Th2) cells 125, 126
television 41, 53, 160
    interventions to limit viewing of 165
Teotihuacan 17
testosterone 82–3
    and immune function in men 82, 96
    levels in western men 94–7
    and muscle mass in men 82
    replacement therapy 117
    variation in levels across populations 95 *see
        also* androgens
thermoregulation *see* energy expenditure
threadworm, *see* helminth infections,
    Enterobius
thriftiness, reconciling the thrifty genotype
    and thrifty phenotype theories 71
thrifty genotype theory 55–61, 65
    applied to Asian populations 160
    and Polynesian migration 57–8
thrifty phenotype theory 68
    applied to Asian populations 160
time, lack of
    as barrier to physical activity 41

Trichuris, see helminth infections
Trowell, Hugh 1
tuberculosis 21
type 1 diabetes
    and breastfeeding 108
    and the hygiene hypothesis 129–30
type 2 diabetes 32, 35–7
    in Asia 159, 160
    and breastfeeding 107
    calpain-10 62
    in children 36
    complications 36
    genes conferring susceptibility to 61–3
    global prevalence 51
    and heart disease 36
    increasing prevalence 36
    in Polynesians 51
    populations with high rates of 50–5
    prevention of 167

urbanisation
    in Africa 161
    and allergy 122, 163
    in Asia 159
    and the burden of non-communicable
        disease 157
    and social isolation 141
Union Army veterans 32, 45–6

walking
    efforts of increase 166
    in hunter–gatherers 28
    in the United States 41
weaning, age at 103
western diseases 6–7
    definition of 1–2
    projected trends 156–63
    rise of 21–2
westernisation
    and asthma 122
    and cardiovascular mortality 48
    definition 1
    and mental health 142
    and stress 151
whipworm, *see* helminths, *Trichuris*
Williams, George C. 2, 4

Zimmet, Paul 37, 52, 58
zoonotic diseases 11